INTERNET POLICIES AND ISSUES

U.S. BROADBAND ISSUES

ADOPTION BARRIERS, PERFORMANCE INFORMATION AVAILABILITY AND MUNICIPAL NETWORKS

INTERNET POLICIES AND ISSUES

Additional books in this series can be found on Nova's website under the Series tab.

Additional e-books in this series can be found on Nova's website under the e-book tab.

INTERNET POLICIES AND ISSUES

U.S. BROADBAND ISSUES

ADOPTION BARRIERS, PERFORMANCE INFORMATION AVAILABILITY AND MUNICIPAL NETWORKS

CHESTER RIOS
EDITOR

nova publishers
New York

Library of Congress Cataloging-in-Publication Data

ISBN: 978-1-63483-944-0

Published by Nova Science Publishers, Inc. † New York

CONTENTS

PREFACE

While broadband is available to a majority of Americans, barriers have kept some from subscribing and enjoying its benefits. In 2010, FCC published the National Broadband Plan, which noted that some demographic groups lagged behind others in adopting broadband and called on FCC and NTIA to take action to address these barriers. This book examines benefits of home broadband adoption; barriers to adoption and approaches to address them; and the extent to which FCC and NTIA have assessed efforts and set goals to address barriers. It also examines broadband performance information available to consumers and its limitations, if any; and FCC's actions to evaluate its efforts to provide consumers with broadband performance information.

In: U.S. Broadband Issues
Editor: Chester Rios

ISBN: 978-1-63483-944-0

Chapter 1

BROADBAND: INTENDED OUTCOMES AND EFFECTIVENESS OF EFFORTS TO ADDRESS ADOPTION BARRIERS ARE UNCLEAR*

United States Government Accountability Office

ABBREVIATIONS

BTOP	Broadband Technology Opportunities Program
CPS	Current Population Survey
DSL	digital subscriber line
FCC	Federal Communications Commission
GPRA	Government Performance and Results Act
Mbps	megabits per second
NTIA	National Telecommunications and Information Administration
Recovery Act	American Recovery and Reinvestment Act of 2009

* This is an edited, reformatted and augmented version of The United States Government Accountability Office publication, GAO-15-473, dated June 2015.

WHY GAO DID THIS STUDY

While broadband is available to a majority of Americans, barriers have kept some from subscribing and enjoying its benefits. In 2010, FCC published the National Broadband Plan, which noted that some demographic groups lagged behind others in adopting broadband and called on FCC and NTIA to take action to address these barriers.

GAO was asked to examine progress in addressing broadband adoption barriers. This report examines (1) benefits of home broadband adoption, (2) barriers to adoption and approaches to address them, and (3) the extent to which FCC and NTIA have assessed efforts and set goals to address barriers. GAO reviewed literature on benefits and barriers, documentation on the performance of efforts to address adoption barriers, and interviewed FCC and NTIA officials, 14 of the 42 BTOP grantees, and 21 public and private stakeholders selected based on GAO's prior work and recommendations from other stakeholders.

WHAT GAO RECOMMENDS

GAO recommends that NTIA include an outcome-based goal and measure for its broadband adoption work in its performance plan. NTIA stated that such metrics are not appropriate for its efforts because these efforts are advisory. GAO believes measuring outcomes is key to demonstrating results. GAO also recommends that FCC revise its strategic plan to more clearly state if broadband adoption is a priority, and if so, what outcomes FCC intends to achieve. FCC noted that to the extent its plan is unclear, it will take steps to address the recommendation.

WHAT GAO FOUND

Home broadband adoption can provide a number of social and economic benefits, according to literature from academic, government, and other research sources and interviews GAO held with researchers, consumer and industry organizations, and government officials. For example, broadband provides access to employment opportunities by providing the means to search and apply for jobs and participate in online job training. It also provides access to a number of government benefits, serves as a conduit for civic participation, and provides a means to connect family members, among other benefits.

Affordability, lack of perceived relevance, and lack of computer skills are the principal barriers to broadband adoption identified by literature and stakeholders GAO interviewed. Efforts to address these barriers include projects to increase broadband adoption that were funded by grants from the National Telecommunications and Information Administration's (NTIA) Broadband Technologies Opportunities Program (BTOP) and outreach and other efforts by the Federal Communications Commission (FCC) and NTIA. GAO identified three key approaches used to address adoption barriers:

- *Discounts* on computer equipment and broadband subscriptions.
- *Outreach* efforts to promote broadband availability and benefits.
- *Training* to help people develop skills in using computers and broadband.

NTIA and FCC have limited information about the performance of their broadband adoption efforts and have not established goals articulating the outcomes these efforts should achieve. For example, NTIA compiled and published self-reported information from its BTOP grantees about best practices, but has not assessed the effectiveness of these approaches in addressing adoption barriers. Because BTOP has concluded, NTIA missed an opportunity to evaluate which grantees' approaches were the most effective. NTIA's strategic plan includes a goal to increase broadband use, but the agency's performance plan does not include an outcome-based goal and performance indicator for its ongoing broadband adoption efforts, making it unclear how the agency will show progress toward its strategic goal. NTIA had reported new broadband subscribers as a result of its BTOP efforts, but no longer uses this as a performance metric because BTOP has largely concluded. Although FCC's previous strategic plan included a goal to support broadband adoption, the commission issued a revised plan in 2015 with fewer broader goals, replacing the goal that mentioned broadband adoption with a goal that instead discusses broadband availability. Although lack of availability is a potential barrier, GAO's literature review and stakeholder interviews more frequently mentioned the three barriers cited above and FCC's broadband adoption efforts are aligned with those barriers. Thus, the strategic plan does not clearly reflect FCC's actions and whether broadband adoption is a priority for the commission. FCC officials said that the new plan's broadband goal is meant to encompass adoption efforts, but without including outcome-oriented goals for broadband adoption, it is unclear what, if any, related outcomes may be expected from FCC's broadband adoption efforts.

* * *

June 2, 2015

The Honorable Frank Pallone, Jr.
Ranking Member
Committee on Energy and Commerce House of Representatives

The Honorable Anna G. Eshoo
Ranking Member
Subcommittee on Communications and Technology
Committee on Energy and Commerce
House of Representatives

The Honorable Edward J. Markey
United States Senate

Internet with speeds at or approaching what is considered high-speed Internet or "broadband" has been deployed to the majority of Americans. According to the Federal Communications Commission (FCC), broadband was available to about 83 percent of U.S. households in 2013, an improvement from about 72 percent in 2011.[1] FCC has also noted, however, that barriers have kept some households from adopting broadband even where it is available, preventing these households from enjoying the social and economic benefits broadband can provide. As required by the American Recovery and Reinvestment Act of 2009 (Recovery Act),[2] FCC released the *National Broadband Plan* in 2010 to improve access to and the services provided by broadband.[3] The plan noted that over one-third of U.S. households lacked broadband and identified affordability, perceived relevance, and lack of computer skills as key barriers to broadband adoption. Additionally, the plan noted that adoption was particularly low among certain demographic segments, including people with low income (less than $20,000 per year), with no high school diploma, age 65 and older, and with a disability. To address barriers to broadband adoption, the plan recommended actions for FCC and the Department of Commerce's National Telecommunications and Information Administration (NTIA). The Recovery Act also provided funds to

NTIA to establish the Broadband Technology Opportunities Program (BTOP). This program awarded about $251 million in grants to 44 projects to improve broadband adoption.[4]

You asked us to review efforts to address broadband adoption barriers described in the *National Broadband Plan*. This report examines (1) what is known about the benefits of home broadband adoption, (2) identified barriers to broadband adoption and approaches used to address them, and (3) the extent to which FCC and NTIA have assessed their effectiveness and set goals for addressing broadband adoption barriers.

To determine what is known about the benefits of home broadband adoption and broadband adoption barriers, we identified and reviewed academic, government, industry and trade press, and other association or nonprofit literature on these topics published from 2009 through 2014.[5] We also interviewed FCC and NTIA officials and 21 stakeholders— including researchers, consumer groups, industry associations and broadband providers, and technology organizations—about broadband benefits and adoption barriers.[6] We identified stakeholders to interview based on our prior telecommunications work and our review of contributors to the *National Broadband Plan* and other broadband adoption literature, as well as based on recommendations from other stakeholders we interviewed. To describe approaches used to address broadband adoption barriers, we identified both federal and nonfederal efforts by reviewing documents from and interviewing FCC and NTIA officials, academic researchers, consumer groups, industry associations, and broadband providers, and by examining 15 of the 44 BTOP projects.[7] We selected these projects because they were the projects that met their goals for new household broadband subscribers as set in their grant applications.[8] For these projects, we reviewed reports submitted to NTIA about project outcomes and, where available, third-party program evaluations that analyzed the effectiveness of approaches to address broadband adoption barriers.[9] We interviewed representatives from the 14 grantees responsible for administering these BTOP projects.[10] To determine the extent to which FCC and NTIA set goals and assessed the agencies' performance to address broadband adoption barriers, we reviewed the agencies' most recent strategic plans and performance plans to identify the agencies' goals and performance measures related to broadband adoption. We reviewed documentation of NTIA's program evaluation of BTOP, including the design and final report. We interviewed FCC and NTIA officials about their efforts to evaluate the effectiveness of their broadband adoption efforts and about their strategic goals and performance measures for these efforts. We evaluated their efforts against criteria for evaluating agency performance as established in our prior work on program evaluation and the GPRA Modernization Act of 2010.[11] To provide contextual information about the level of broadband adoption among households overall and within certain demographic groups, we analyzed U.S. Census Bureau data from 2009 through 2013. We determined that these data were sufficiently reliable for showing differences among demographic groups and trends by reviewing information about how Census and NTIA have reported these data and interviewing Census and NTIA officials about these data and their limitations. See appendix I for a more detailed description of our objectives, scope, and methodology.

We conducted this performance audit from August 2014 to June 2015 in accordance with generally accepted government auditing standards. Those standards require that we plan and perform the audit to obtain sufficient, appropriate evidence to provide a reasonable basis for our findings and conclusions based on our audit objectives. We believe that the evidence obtained provides a reasonable basis for our findings and conclusions based on our audit objectives.

BACKGROUND

The Recovery Act: FCC and NTIA Broadband Roles

Most of the *National Broadband Plan*'s recommendations on broadband adoption were made to either FCC or NTIA, two federal agencies with complementary responsibilities for broadband. FCC regulates wired and wireless forms of communication, including the means of providing broadband service, such as cable, telephone, and satellite technologies. FCC's role includes promoting competition, protecting consumers, and addressing the communications needs of underserved populations. For example, FCC's Lifeline Program provides a subsidy to providers that offer discounted telephone service to low-income households. Through a broadband pilot, FCC has been examining ways to expand the Lifeline program to provide support for broadband service.[12] NTIA is part of the Department of Commerce and advises the President on telecommunications and information-policy issues, including broadband access and adoption. NTIA's activities include administering grant programs that further the deployment and use of broadband and other technologies and developing policy on issues related to the Internet economy, including online privacy, copyright protection, and cybersecurity.

The Recovery Act included provisions aimed at FCC and NTIA to address broadband adoption. The act provided $4.7 billion to NTIA to establish BTOP, through which NTIA awarded competitive grants to a variety of entities for broadband infrastructure, public computer centers, and to increase broadband access and adoption. Through the broadband adoption component of the program—known as the Sustainable Broadband Adoption grant program—NTIA awarded about $251 million to fund 44 broadband adoption projects across the country. [13] The purpose of this program was to increase Internet use and broadband subscribership among individuals and businesses. The Recovery Act also authorized NTIA to transfer a portion of its funds to FCC to develop the *National Broadband Plan*, a document to outline ways to ensure that every American had access to broadband service. FCC published the *National Broadband Plan* in 2010.[14] The plan's chapter on broadband adoption provided data on the status of broadband adoption among U.S. households overall and within certain demographics, discussed the barriers to adoption faced by those that lack broadband at home, and recommended actions for FCC and NTIA to address those barriers.[15]

Trends in Household Broadband Adoption

According to U.S. Census data, an estimated 73 percent of U.S. households had adopted broadband as of 2013, a statistically significant increase of 9 percentage points from 2009.[16] Estimated broadband adoption levels were significantly higher in 2013 than 2009 for all of the demographic groups we examined; however, some groups had estimated levels of broadband adoption below the level for households overall. For example, the estimated adoption level among households headed by a person age 65 and older increased a statistically significant 14 percentage points from 40 percent to 54 percent from 2009 to 2013.[17] Hispanics also showed a similar increase during this time—an estimated 12 percentage points, from 49 to 61 percent.[18] However, older Americans and Hispanics are an estimated 19 and 12 percentage points, respectively, below the adoption rate for all households. These and other demographic groups, including those with less than a high school education, African Americans, and those with a disability, showed similar increases while still falling behind households overall, as shown in figure 1.

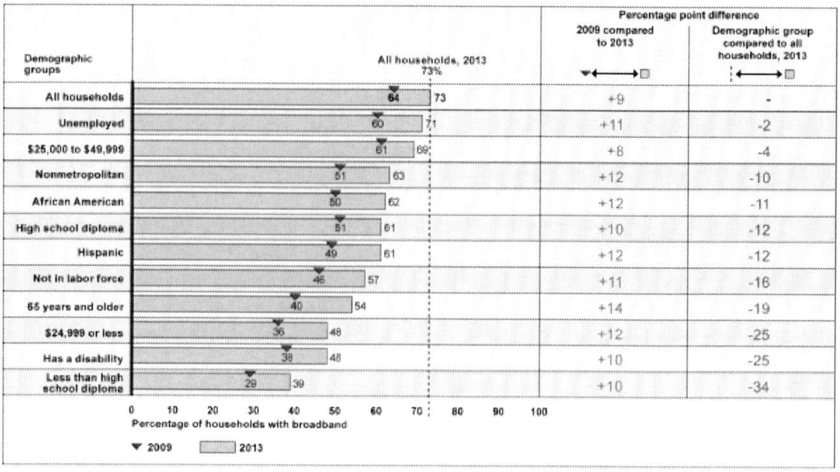

Source: US Census Current Population Survey and GAO analysis. | GAO-15-473.

Note: The Census Current Population Survey asks the respondent if anyone in the household uses the Internet from home and then asks through which types of service the Internet is accessed. Respondents who say they use dial-up are not counted as having broadband. Respondents who indicate they access the Internet through DSL, cable modem, fiber-optic service, mobile broadband plan, satellite, or some other service are counted as a household with broadband. The household broadband adoption percentages we present are estimates calculated with intervals of 95 percent confidence. They have a sampling error of no larger than plus or minus 2.6 percentage points at the 95 percent level of confidence. Where we have drawn comparisons between groups or years to show differences, those differences are statistically significant and the confidence intervals do not overlap, except in the 2013 comparison between all households (95 percent confidence interval of 72.70 to 73.69 percent) and unemployed (95 percent confidence interval of 68.36 to 73.58 percent). About 1.6 percent of household respondents indicated they were more than one race; we have included these householders under multiple demographic race categories. Starting in 2010, Census changed the methodology for its income variable, including imputing household values that had been missing in prior years to provide values without missing observations.

Figure 1. Household Broadband Adoption among Demographic Groups with Adoption Rates below That of All Households, 2009 and 2013.

There are also some differences in broadband adoption geographically. For example, nonmetropolitan households' 2013 estimated adoption rate of 63 percent is 12 percentage points less than the estimated 75 percent adoption rate of metropolitan areas.[19] Broadband adoption in the states also varies. According to U.S. Census's American Community Survey data, it is estimated that broadband adoption in Alabama, Arkansas, Mississippi, New Mexico, and Puerto Rico was nearly 10 percentage points lower than the national average in 2013.[20] See appendix II for more information on broadband adoption trends.

HOME BROADBAND OFFERS A VARIETY OF ECONOMIC AND OTHER BENEFITS

Adopting broadband at home—as opposed to through mobile devices—can provide a number of benefits, including access to job opportunities, education, and services for economic and social gain.[21] Although mobile devices are becoming more advanced, stakeholders told us that certain benefits are not as easily accessible through mobile devices. For example, it can be difficult, if not impossible, to use a mobile device to develop and submit a resume along with a job application. Additionally, some benefits may also be accessed through broadband connections in public places, such as libraries. However, such access also has limitations, such as time limits. As stated in the *National Broadband Plan*, not adopting broadband limits individuals' ability to fully participate in digital information and services. The literature we reviewed and stakeholders we interviewed identified a variety of benefits:

- *Employment opportunities:* Broadband provides a mechanism to search and apply for jobs. Many employers now require applications to be completed online, including the federal government. In 2011, FCC noted that over 80 percent of Fortune 500 companies require applicants to apply for employment on the Internet.[22] Telecommuting has also allowed many government and private sector employees (e.g., including individuals with disabilities) to use broadband to work at home.
- *Educational enrichment:* Broadband provides an opportunity for web-based learning, for example to complete college requirements or job training and for school-age children to use the Internet for research and classroom assignments. NTIA's BTOP evaluation contractor noted in its study design document that prior research shows using computers and broadband can be a significant factor in boosting math and reading achievement as well as motivating students and enriching school content, ultimately improving students' achievement.[23]

- *Healthcare:* Broadband provides consumers the ability to research health issues, obtain and share their personal health information with third parties, and to communicate with doctors, including specialists who may work in a different city. For example, in 2010, we found that through remote access, telemedicine[24] can allow rural patients to receive medical diagnosis or patient care, including from specialists who are located elsewhere.[25]
- *Social communication and networking:* Broadband provides a mechanism for people to communicate with friends, family members, and organizations. For example, it provides an opportunity for older adults who may be socially isolated to participate in social dialogue about current events and issues.
- *Government services:* Broadband provides an opportunity to obtain information about and apply for most government entitlement benefits, such as social security, and to complete tasks such as tax filing.
- *Electronic commerce:* Broadband allows online sales transactions. Consumer studies and stakeholders we interviewed reported that the Internet has become an important part of the U.S. economy. This is also reflected in our prior work. In 2014, we found that electronic commerce provides opportunities for local residents to shop online rather than at brick-and-mortar establishments.[26] Shopping online also enables comparing prices and searching for price discounts, thus lowering the price paid for goods and services.
- *News and civic engagement:* Broadband provides a variety of sources for national and local news. In 2013, Pew Research Center reported that 82 percent of Americans said they accessed news with a desktop or laptop computer.[27] Broadband can also enhance household participation in various government activities that will promote interaction between citizens and their governments.
- *Entertainment:* Broadband provides the ability to access games, movies, music, books, and other forms of entertainment. For example, movies can be streamed and watched from a computer or television.

Additional benefits may emerge as more information and services become available through broadband. The *National Broadband Plan* noted that many uncertainties will shape the evolution of broadband, including the behavior of consumers, private companies, the economic environment, and technological advances.

Affordability, Perceived Relevance, and Lack of Computer Skills Are Identified Broadband Adoption Barriers, Which Discounts, Outreach, and Training Have Sought to Address

Affordability, Perceived Relevance, and Lack of Computer Skills Are Identified Broadband Adoption Barriers

In 2010, FCC's *National Broadband Plan* identified affordability, perceived relevance, lack of computer skills, and the availability of broadband service as the four principal barriers to broadband adoption.[28] Our more recent literature review and interviews with stakeholders including FCC and NTIA officials, researchers, industry associations and companies, and consumer groups showed that three of these barriers persist:

- Affordability: All 21 stakeholders we interviewed identified affordability issues, such as the cost of a subscription for Internet service and computer equipment, as barriers to broadband adoption. While stakeholders told us that it is difficult to find data on broadband subscription costs, affordability was the most frequently identified barrier in the literature and among the stakeholders we interviewed. [29]
- Perceived relevance: Almost all of the stakeholders we spoke to told us that the perception that broadband does not provide enough utility, relative to its cost, acts as a barrier to broadband adoption. A 2013 NTIA report noted this was the most-cited reason households had not adopted broadband.[30] While some individuals may not benefit from broadband adoption, the literature we reviewed and stakeholders we interviewed told us that many people who have not adopted broadband are not aware of its benefits. Other underlying factors may affect the perception of relevance. For example, NTIA officials and other stakeholders said that some people who say that they do not consider broadband to be relevant may in fact be unable to afford it or lack the computer skills necessary to use it.
- *Lack of computer skills.* A majority of the literature we reviewed and the stakeholders we interviewed told us that lack of computer and Internet skills is a barrier to broadband adoption. Stakeholders we interviewed said that this is a particular problem for certain demographic segments that may lack exposure to or knowledge about

computers, such as those of age 65 and older and those with low levels of income and education.

The literature we reviewed and stakeholders we spoke to said that the barrier of availability—the inability to access broadband because it is not offered in a given area by a provider—has been addressed throughout most of the country, but remains an issue primarily for rural areas. Depending on what speed metric is used, FCC's broadband availability data show that broadband was available to all but 4 to 17 percent of the population in 2013.[31] These data also show some geographic variation in the availability of broadband, specifically that broadband was unavailable to 20 to 53 percent of the rural population, as well as 17 to 63 percent of Americans living on tribal lands.

Various types of stakeholders we interviewed also identified other barriers to broadband adoption in addition to those identified in the *National Broadband Plan*, although with less frequency:

- *Language:* Six stakeholders told us that language issues—either lack of English fluency among some immigrants or illiteracy among some low-income households—are barriers.
- *Data security concerns:* Three of the 21 stakeholders we spoke with identified data security concerns such as privacy and identity theft as a barrier to broadband adoption.
- *Accessibility for people with disabilities:* NTIA and FCC, as well as 2 of the 21 stakeholders we interviewed identified accessibility as an issue for people with disabilities, such as those who are blind, deaf, or hard of hearing.

Approaches to Address Broadband Adoption Barriers Include Discounts, Outreach, and Training

Our review of projects funded by NTIA's BTOP, other FCC and NTIA efforts, efforts by the private sector and nonprofit organizations, and interviews with stakeholders identified three primary approaches that have been used to address broadband adoption barriers:

- *Free or discounted equipment and broadband service:* Broadband adoption efforts may provide computer equipment or broadband service at reduced or no cost. This approach addresses the barrier of affordability.

- *Public outreach:* Outreach campaigns seek to promote broadband adoption programs or improve awareness of broadband and its benefits. Outreach efforts may include partnering with community organizations, which representatives of some BTOP projects told us was an effective way to address issues of trust and relevance among lower-income populations that are sometimes suspicious of such outreach. This approach addresses the barrier of relevance.
- *Training (computer or Internet use):* Training may focus on basic computer skills such as operating a computer, using e-mail, and navigating the Internet, or more advanced and specialized skills. This approach addresses the barrier of the lack of computer skills.

Table 1 shows examples of efforts under these three primary approaches.

Table 1. Examples of Efforts to Address Barriers to Broadband Adoption

Approach	Example efforts
Free or discounted equipment or broadband service	• The Tampa Housing Authority, a Broadband Technology Opportunities Program (BTOP) grantee, provided subsidized computer equipment and broadband service in public-housing units. • Comcast's Internet Essentials and CenturyLink's Internet Basics are programs that offer discounted broadband service and computer equipment to qualified low-income households.
Public outreach	• EveryoneOnand Connected Nation are technology organizations that work to establish partnerships with Internet service providers or state or local governments. Such partnerships promote the benefits of broadband adoption or provide discounted broadband service to low-income families. • Using BTOP funds, the California Emerging Technology Fund conducted a public awareness campaign promoting the benefits of broadband adoption in multiple languages.
Training (computer or Internet use)	• The Mexican Institute of Greater Houston, a BTOP grantee, offered classroom instruction for adultson basic computer literacy in English and Spanish. • The City and County of San Francisco used its BTOP grant to

Table 1. (Continued)

Approach	Example efforts
	partner with media-focused non-profit organizations to establish computer centers in city schools that offered digital media training to students.

Source: GAO analysis of BTOP projects and interviews with stakeholders. | GAO-15-473.

Note: These are examples of efforts that align with the various approaches; organizations cited may also have other broadband adoption efforts that align with other approaches.

We identified efforts incorporating these approaches that generally targeted specific demographic groups. The majority of federal efforts we examined were projects that received grants through NTIA's BTOP program, which was funded through the Recovery Act, but do not receive federal funds on an ongoing basis. Additionally, both NTIA and FCC had some additional efforts. We examined the following efforts:[32]

- *BTOP broadband adoption projects:* As previously discussed, NTIA funded these 44 projects through a grant program established by the Recovery Act. These projects were implemented in various states and targeted multiple demographic groups. Grantees included state and city government agencies, universities, public-private partnerships, schools, libraries, and nonprofit organizations. BTOP grantees frequently used combinations of the three approaches to addressing barriers to broadband adoption described above. For example, Connected Tennessee partnered with a Tennessee state agency to distribute laptop computers to children leaving the foster care system. Connected Tennessee also partnered with a youth services organization to establish computer centers offering training to children. Although most of the BTOP funds have been spent, some projects have continued with other sources of funding.[33] For example, the Mexican Institute of Greater Houston has continued to operate its computer-training classes with funding from Houston-area donors.
- *Digitalliteracy.gov:* NTIA operates a website to provide resources for teaching computer and Internet skills, as well as tools to aid in finding job opportunities and career training services.

- *FCC's Lifeline broadband pilot:* FCC pilot tested ways to provide subsidized broadband service through its Lifeline program, which provides subsidies to telecommunications providers to offer discounted telephone service to low-income households. We examined this program in a recent report.[34]
- *FCC's National Deaf-Blind Equipment Distribution Program:* This program provides $10 million annually for the local distribution of accessible communications equipment for use with broadband services (as well as other telecommunications services) to low-income individuals who are deaf and blind.
- *Other outreach efforts by NTIA and FCC:* These agencies have conducted other outreach efforts, such as using public-private partnerships to promote broadband adoption. They have also held meetings, workshops, and conferences dedicated to promoting broadband adoption. These outreach efforts generally target specific demographic groups. For example, FCC officials told us they have participated in events intended to increase broadband use among older adults, including events hosted by AARP. FCC also noted taking a number of actions to address coordination with Tribal Nations, including creating an Office of Native Affairs and Policy in response to a *National Broadband Plan* recommendation to ensure effective coordination and consultation with tribes on broadband-related issues.[35] According to NTIA officials, upcoming outreach efforts include providing technical assistance to at least 175 communities in implementing or expanding broadband availability and adoption programs.
- *Broadband provider discounts*: Some Internet service providers offer reduced-price broadband service to qualifying lower-income families. For example, the two companies we met with offer monthly service for about $10 to families with children enrolled in federally subsidized school lunch programs. Such discounts may be supplemented with other approaches to encourage adoption, such as training on how to use the Internet.
- *Technology organizations:* These organizations work to establish partnerships with Internet service providers, state and local government agencies, and other organizations. These partnerships encourage broadband adoption through efforts such as promoting the benefits of broadband and helping low-income households obtain discounted Internet subscriptions. They may also engage in other

efforts, such as helping individuals find free computer skills training in their area or providing computer equipment to low-income populations.

NTIA IDENTIFIED BEST PRACTICES FOR BROADBAND ADOPTION, BUT NTIA AND FCC HAVE NOT FULLY ASSESSED THE EFFECTIVENESS NOR CLEARLY ARTICULATED INTENDED OUTCOMES OF THEIR BROADBAND ADOPTION EFFORTS

NTIA Identified Best Practices of BTOP Projects, but Evaluation of the Effectiveness of NTIA's and FCC's Efforts Is Limited

NTIA has collected and disseminated information about the practices used by the projects funded by its BTOP grant program to address barriers to broadband adoption. NTIA officials told us they intentionally selected a diverse group of 44 projects to receive BTOP funding in order to discover best practices for encouraging broadband adoption. Grantees provided this information on their practices to NTIA as part of their required progress reporting,[36] and NTIA also contracted with a third-party, ASR Analytics, to conduct a program evaluation of the BTOP program as a whole and report its results.[37] NTIA compiled grantee-reported best practices and published them as a broadband adoption toolkit in 2013.[38] The practices represent a variety of strategies to address broadband adoption barriers that align with the three approaches we described earlier in this report, such as strategies for reaching out to different demographic groups and providing various types of computer and Internet training. NTIA also monitored the projects' progress against their stated goals, including the number of new broadband subscribers as a result of the project and the status of efforts such as outreach and training programs. NTIA's evaluation contractor, ASR Analytics, submitted its evaluation report to NTIA in September 2014, and NTIA later made the report available on its website.[39] The report includes a section on best practices similar to the information in NTIA's toolkit, such as best practices for outreach and communication, instruction, and ways to leverage and sustain community partnerships.

While NTIA's and ASR's reports summarize BTOP project practices, neither evaluated these practices to examine their effectiveness at addressing broadband adoption barriers. NTIA's toolkit links grantees' identified best practices to specific populations and adoption barriers they are meant to address; however, NTIA officials said they did not develop criteria for assessing the success of these practices in getting new broadband subscribers. Additionally, the focus of the ASR evaluation was to identify social and economic outcomes, not which approaches for addressing broadband adoption barriers were the most effective. NTIA officials told us they have not evaluated the effectiveness of their other broadband adoption efforts—the digital literacy website and their ongoing outreach efforts. NTIA officials said that evaluating their efforts to address broadband adoption barriers is challenging and potentially expensive, and therefore the agency has not pursued doing so. They also said that assessing the impact of these projects on broadband adoption is difficult given the external social factors that also affect broadband adoption. Further, the officials told us that the motives for why people do or do not adopt broadband are well understood.

Some BTOP grantees had third-parties evaluate their broadband adoption projects. Specifically, among the 15 projects we examined, 3 were evaluated by third parties seeking to identify the effectiveness of their approaches for addressing adoption barriers.[40] These evaluations took different forms. For example, a team of academics evaluated the effectiveness of the Chicago Smart Communities initiative by comparing broadband adoption data in communities where the project operated to similar communities where it did not. The evaluation found that broadband adoption was significantly higher in neighborhoods where the program was active than in similar neighborhoods where it was not. Researchers suggested that participants' increased experience using the Internet and the introduction of low-cost offerings from a broadband provider were possible explanations for the increase despite the program's not offering discounted service.

FCC plans to evaluate its Lifeline broadband pilot, but FCC officials told us they have not evaluated the effectiveness of the commission's other efforts to address broadband adoption barriers. Our March 2015 report examining the Lifeline program found weaknesses in FCC's ability to effectively evaluate the program. Specifically, we found that the usefulness of information FCC gathered through the broadband pilot program may be limited due to a lack of an evaluation plan and other challenges.[41] Similar to what NTIA officials told us, FCC officials said that evaluating the impact of their efforts to address broadband adoption barriers would be complex and potentially expensive.

Program evaluation is a useful tool for understanding the performance of government programs. It can be used to systematically determine the "value added" by the expenditure of public resources. Specifically, program evaluation can validate which approaches are effective, measure results that are difficult or expensive to assess, and explore why performance goals are not met.[42] Our prior work has shown that program evaluation can help agencies determine their program performance, assess program effectiveness, and contribute to improving program management.[43] The *National Broadband Plan* discussed the need for program evaluation of broadband adoption programs, stating that such studies are necessary for understanding costs, benefits, and efficiency of different adoption programs, without which the country has a limited understanding of which approaches work.

As we discussed earlier, FCC and NTIA have undertaken a number of initiatives aimed at addressing the barriers to broadband adoption, but since evaluation of these efforts has been limited, the agencies have limited information about which approaches are the most effective. For example, the BTOP project practices NTIA identified may hold promise for addressing broadband adoption barriers, but it is unclear how effective those practices are. Out of the 43 BTOP broadband adoption projects that were completed, more than half did not meet their goals for the number of new household broadband subscribers.[44] Further, among the 24 BTOP broadband adoption grantees NTIA included in its toolkit of promising strategies, 12 did not meet their subscriber goals. While other factors may have contributed to these projects not meeting these goals, without more rigorous evaluation, the ability to determine whether the approaches did or did not work well or if other factors contributed to their not meeting these goals is limited. Furthermore, it is not clear in which areas the agencies should dedicate scarce federal resources to have the greatest impact in addressing barriers to broadband adoption. Because NTIA's major broadband adoption effort—BTOP—is substantially complete, and FCC ended the subsidy for the Lifeline broadband pilot in October 2014, the agencies missed opportunities to incorporate program evaluation into the design of their broadband adoption efforts in order to assess the effectiveness of these efforts in addressing barriers to broadband adoption.

FCC's Strategic Plan and NTIA's Performance Plan Do Not Clearly Convey Quantifiable Outcomes of the Agencies' Efforts to Address Broadband Adoption Barriers

FCC and NTIA both have ongoing efforts to address broadband adoption barriers, but it is unclear what outcomes they intend to achieve with these efforts because FCC's strategic plan and NTIA's performance plan do not clearly communicate the agencies' desired outcomes for their efforts to address broadband adoption barriers. Communicating what an agency intends to achieve and its progress in doing so are fundamental aims of performance management. Under the GPRA Modernization Act of 2010, an agency is to clearly communicate the intended outcomes of its efforts with a strategic plan and annual performance plans. Specifically, the GPRA Modernization Act of 2010 requires an agency to have:

1. A strategic plan containing outcome-oriented goals for major functions and operations and a description of how goals are to be achieved.
2. An annual performance plan consistent with the strategic plan with quantifiable, measurable performance goals to define the level of performance to be achieved during the year and performance indicators to assess progress toward each performance goal.

The *National Broadband Plan* made recommendations to FCC and NTIA for actions to address broadband adoption barriers. While officials from both agencies told us they have taken steps consistent with recommendations in the plan, they indicated that the plan is not the basis of their ongoing strategies. FCC officials told us that the commission has taken steps to implement actions consistent with a number of the plan's recommendations. However, they also said that the National Broadband Plan was intended to be an evolving document and that FCC continues to evaluate the relevance of its recommendations as technologies, markets, and consumer needs and behaviors change. NTIA officials told us that their agency provided input for the National Broadband Plan based on activities already underway within the agency. Specifically, NTIA officials told us that the initial steps to implement BTOP were taken in 2009 before publication of the plan in early 2010. Thus, officials told us that NTIA's efforts in this area were guided more by perceived needs than the National Broadband Plan. Further, NTIA officials told us that no funding was made available for the agency to implement recommendations in the plan.

Nonetheless, both agencies remain involved in addressing broadband adoption barriers. Although some of their recent efforts have concluded, a number of FCC and NTIA efforts are ongoing, as we discussed earlier in this report. NTIA's strategic plan makes it clear that addressing broadband adoption barriers is a major function of the agency. Broadband adoption is mentioned prominently and early in NTIA's fiscal year 2016 budget submission to Congress and in the description of the agency's purpose on its website.[45] FCC includes updates on broadband adoption in its broadband progress reports. However, what outcomes can be expected with these efforts is unclear because FCC's strategic plan and NTIA's performance plan do not clearly communicate what they intend to achieve:

- *FCC:* In 2015, FCC released an updated strategic plan for fiscal years 2015–2018 with a strategic goal of "making networks work for everyone" including ensuring that "all Americans can take advantage of the services they provide without artificial impediments." This goal has a supporting strategic objective to "maximize availability of broadband to all." FCC's prior strategic plan, covering fiscal years 2014–2018, included a strategic objective to "maximize" broadband adoption and a related performance goal to "support and facilitate" broadband adoption. Although the prior strategic plan made it clear that addressing broadband adoption was a priority of FCC, the revised plan's focus on availability makes this unclear. As we discussed earlier in this report, while availability may be a barrier for some— particularly in rural areas—other issues, such as affordability, relevance, and computer skills are the principal barriers to adoption that the literature we reviewed and stakeholders we spoke to identified. Further, the efforts FCC officials told us comprise their work to address broadband adoption barriers target populations affected by these three principal barriers, such as their efforts to help low-income households afford broadband services and equipment or promoting broadband to older individuals and those living on tribal lands. FCC officials told us that the new plan was designed to have fewer but broader strategic goals and that the goals and objectives of the new plan are meant to support broadband adoption by promoting the availability of broadband, ensuring schools and libraries have broadband access, and ensuring that Americans with disabilities can access broadband—activities that align in part but not completely with FCC's work to address adoption barriers. The goals of an agency's strategic plan are supposed to express intended outcomes and link to major agency functions consistent with the GPRA Modernization Act of 2010. With the changes in FCC's strategic plan to focus on availability, rather than adoption, it is unclear whether broadband adoption is a major function and thus a current commission priority and what outcomes can be expected of FCC's ongoing broadband adoption efforts.

- *NTIA:* The Department of Commerce's 2014–2018 strategic plan includes a goal to foster a more innovative economy under which NTIA has a key strategy to increase broadband infrastructure and use, indicating that broadband adoption is an NTIA priority. However, now that BTOP has concluded, NTIA no longer has a quantifiable performance goal or measure that assesses outcomes consistent with the GPRA Modernization Act of 2010. The Department of Commerce's Budget in Brief for fiscal year 2016 states that NTIA will work to increase broadband adoption through technical assistance and includes the number of communities in which NTIA has provided technical assistance as a performance measure.[46] Because this measure articulates an activity of the agency, rather than a result of an agency activity, it measures an output and not an outcome. According to NTIA officials, the agency's broadband adoption goals reflect what the agency has been funded to do and beyond BTOP, the agency has received limited funding for broadband adoption efforts. NTIA's performance plan used to include an outcome-oriented performance measure: the cumulative number of new broadband subscribers as a result of BTOP broadband adoption projects. However, the fiscal year 2016 budget states that this indicator has been retired as of fiscal year 2015 because NTIA has met its final target. NTIA officials also told us that is the case because BTOP has concluded. The budget request does seek funding for the outreach and other ongoing efforts we described earlier in this report, indicating that NTIA does intend to move forward with other broadband adoption efforts.[47]

CONCLUSION

Without an outcome-oriented goal for increasing broadband adoption in its strategic plan, it is unclear whether this is still a priority for FCC, as well as what the commission intends to achieve with its ongoing broadband adoption efforts. Although NTIA's strategic plan more clearly outlines broadband adoption as an agency priority, without goals or measures in its performance plan that point to outcomes, it is unclear what NTIA intends achieve with its post-BTOP broadband adoption efforts and how it will show progress in achieving its related goals.

The *National Broadband Plan* laid out a vision for why increasing broadband adoption is important and steps the federal government— namely FCC and NTIA—could take to achieve that vision. In the ensuing years, adoption has increased, but a significant percentage of the population has still not adopted broadband, and non-adoption rates remain higher among populations such as low-income households and older Americans. The various ways that people use broadband for everyday social and economic purposes underscore that the benefits of broadband are substantial. Through BTOP, Lifeline, and other initiatives, NTIA and FCC have worked to address barriers to broadband, but the agencies' current goals and the intended outcomes of their work are unclear. NTIA's investment in BTOP broadband adoption projects in particular represented a significant effort to address these barriers, but with that program coming to an end, ongoing efforts to address these barriers are limited. For both of these agencies, key planning documents released in 2015 that are to clearly communicate intended outcomes have dropped or revised goals included in prior years' plans related to the agencies' efforts to address broadband adoption. Consequently, it is unclear that the agencies have a clear strategy for sustaining their efforts to increase broadband adoption, and the ability to tell whether these efforts will be successful is limited.

RECOMMENDATIONS

To more clearly establish the outcomes FCC intends to achieve through its efforts to address broadband adoption barriers faced by demographic groups with low levels of adoption, we recommend that FCC take the following action:

- revise its strategic plan to more clearly indicate whether addressing broadband adoption barriers is a major function, and if so, specify what outcomes the commission intends to achieve.

To more clearly communicate what NTIA hopes to achieve with its ongoing broadband adoption efforts and to demonstrate the agency's progress, the Secretary of Commerce should take the following action:

- include in NTIA's annual performance plan a quantifiable outcome-based performance goal and a related performance indicator for the agency's broadband adoption efforts consistent with the strategic plan.

AGENCY COMMENTS AND OUR EVALUATION

We provided a draft of this report to FCC and the Department of Commerce for comment. Neither agency specifically agreed or disagreed with our recommendations but provided comments as summarized below.

In response to our recommendation, FCC's comments state that broadband adoption remains a significant focus for the commission. Further, FCC noted that in its revised strategic plan, it presented broader goals to achieve a more streamlined format but that its strategic objectives, performance goals, and two of the three strategies associated with the plan's Strategic Goal 3: Making Networks Work for Everyone are also directed at promoting adoption. However, FCC states that to the extent that its strategic plan is unclear on the role of broadband adoption in FCC's efforts, it is prepared to implement revisions that will help clarify this issue. With regard to quantifying expected outcomes of FCC's efforts in this area, FCC notes that it intends to seek comments on this issue as part of a proposed rulemaking for the Lifeline program. Specifically, the response states that FCC has drafted a Further Notice of Proposed Rulemaking seeking to refocus the program on "21st century services." FCC stated this notice will seek comment on how to best consider specific outcomes related to addressing the barrier of affordability. FCC notes it has also hired a new staff person to lead an effort to address other broadband adoption barriers, including lack of computers or perceived relevance. FCC also provided technical comments, which we incorporated as appropriate.

In response to our recommendation, the Department of Commerce states that it agrees that outcome-based performance tools are important in evaluating the success of an agency program. However, it also states that an outcome-based measure like the one it used for BTOP would not be an accurate measure of NTIA's performance because the agency's ongoing efforts to provide technical assistance to communities to support broadband adoption efforts are advisory and do not provide funding to communities. Additionally, Commerce states that the communities identify the goals of these efforts. Consequently, Commerce believes other variables unique to these communities are more appropriate for determining whether a community adopts broadband or not. Further, Commerce states that NTIA's fiscal year 2015 budget allocated $3 million for its broadband initiative, but none of these funds are authorized to be used for grants to establish broadband adoption programs. However, Commerce states that it would seek to develop outcome-based measures if it received funding for broadband adoption grants.

Our prior work has stressed the importance of setting goals and measuring progress toward those goals. We have acknowledged that agencies sometimes face difficulties in developing and using outcome measures, such as accounting for factors that are outside of an agency's control.[48] However, we have also found that outcome-oriented performance measures are important for helping agencies determine if their efforts are achieving their intended purpose. This includes grant programs, like BTOP, but other types of federal efforts also measure outcomes to demonstrate performance. NTIA has experience with measuring outcomes under challenging circumstances when it implemented its broadband adoption effort under BTOP, which consisted of over 40 projects in varying communities throughout the country. NTIA officials said that measuring broadband adoption was challenging and that grantees employed different methodologies to measure the number of new broadband subscriptions. NTIA officials said they worked with BTOP grantees to develop appropriate and consistent methodologies for measuring the performance of their projects. Given NTIA's experience in working with BTOP grantees to develop ways to measure broadband adoption, we believe the agency could apply what it learned from that effort to assess outcomes for its ongoing broadband adoption efforts. As our report notes, without goals or measures in its performance plan that point to outcomes, it is unclear what NTIA intends achieve with its ongoing broadband adoption efforts and how it will show progress in achieving its related goals. Thus, we continue to believe that our recommendation has merit and should be fully implemented.

Mark L. Goldstein
Director,
Physical Infrastructure Issues

APPENDIX I: OBJECTIVES, SCOPE AND METHODOLOGY

This report examines (1) what is known about the benefits of home broadband adoption, (2) identified barriers to broadband adoption and approaches used to address them, and (3) the extent to which the Federal Communications Commission (FCC) and National Telecommunications and Information Administration (NTIA) have assessed their effectiveness and set goals for addressing broadband adoption barriers.

To determine what is known about the benefits of home broadband adoption, we identified and reviewed academic, government, industry, trade press, and other association or nonprofit literature on broadband benefits. To identify this literature, we conducted searches of various databases such as ProQuest, Lexis.com, and PolicyFile. We included in our review additional reports we obtained from NTIA. We limited the scope of literature to focus on (1) broadband benefits specific to home broadband, as opposed to mobile broadband; (2) experience within the United States; and (3) articles from January 2009 to November 2014, a timeframe we chose because it represents recent years just before and since the 2010 publication of FCC's *National Broadband Plan*. We also interviewed FCC and NTIA officials and 21 stakeholders—including researchers, consumer groups, industry associations and broadband providers, and technology organizations—about broadband benefits (see table 2 for a list of stakeholders interviewed). We identified these stakeholders as having broadband expertise based on our prior telecommunications work, their contributions to the *National Broadband Plan* and other broadband adoption literature, and recommendations from these stakeholders.

For contextual purposes, we analyzed household broadband adoption data from the U.S. Census Current Population Survey (CPS) for 2009 through 2013.[1] We used variables included in the survey for our household demographic estimates. Because the CPS uses a complex survey design, all estimates derived from CPS data have sampling error associated with them. We used the sample weights provided in the CPS data and accounted for the design of the survey in our estimation of the sampling error. We express our confidence in the precision of estimates as 95 percent confidence intervals (e.g., plus or minus 2 percentage points). This is the interval that would contain the actual population value for 95 percent of the CPS samples the Bureau of the Census could have drawn.

Table 2. Academic, Consumer, Industry, and Technology Stakeholders GAO Interviewed

Stakeholder type	Name
Researchers	John Horrigan, Pew Research Center Blair Levin, Brookings Institute Jason Llorenz, Rutgers University
Consumer organizations	AARP American Association of People with Disabilities American Library Association[a] Common Cause[b] Consumer Federation of America Multicultural Media, Telecom and Internet Council National Hispanic Media Coalitionb National Urban League[b] New America Foundation Older Adults Technology Services United Church of Christb
Industry organizations and companies	CenturyLink Comcast National Cable and Telecommunications Association NTCA – The Rural Broadband Association US Telecom
Technology organizations	Connected Nation Everyone On

Source: GAO. | GAO-15-473.

[a] We have categorized the American Library Association as a consumer organization because the focus of our discussion was how broadband adoption barriers faced by consumers impact libraries.

[b] We conducted an interview with these organizations jointly, the result of outreach to the National Consumer Law Center. Representatives from the National Consumer Law Center were unable to participate in the meeting.

For the various comparisons used in our analysis of the CPS data, we determined that two estimates were statistically significant when the 95 percent confidence intervals around the two estimates did not overlap. To assess the reliability of these data, we reviewed how these data have been reported by Census and NTIA and discussed with Census and NTIA officials reasons for differences in how they have reported these data, any potential limitations with how we planned to use CPS data in our report, and trade-offs between using that and another Census product, the American Community Survey, which also includes questions about broadband adoption. We determined that these data were sufficiently reliable for the purposes of showing demographic differences and recent trends in household broadband adoption.

To identify broadband adoption barriers, we identified and reviewed academic, government, industry and trade press, and other association or nonprofit literature published from January 2009 to November 2014. We also interviewed academic researchers, consumer groups, industry associations and companies, and FCC and NTIA officials about broadband adoption barriers (see previous table 2 for a list of stakeholders interviewed). To describe federal and nonfederal efforts to address broadband adoption barriers, we identified efforts by reviewing documentation and interviewing researchers, consumer groups, industry associations and companies, and FCC and NTIA officials.[2] For additional information about potentially promising approaches to addressing broadband adoption barriers, we selected Broadband Technology Opportunities Program (BTOP) Sustainable Broadband Adoption grant projects that met their goals for new broadband subscribers as set in their grant applications, which was 15 of the 44 projects. For these projects, we reviewed projects' reports submitted to NTIA about project outcomes, and, where available, third-party program evaluations that analyzed the effectiveness of their approaches to address broadband adoption barriers.[3] We interviewed representatives from the 14 grantees responsible for administering these 15 BTOP projects, which are listed in table 3.

Table 3. Selected BTOP Projects Grantee (state)

Grantee (state)	Project(s)
C.K. Blandin Foundation (Minnesota)	Minnesota Intelligent Rural Communities
California Emerging Technologies Fund (California)	Broadband Awareness and Adoption; Access to Careers and Technology
City and County of San Francisco (California)	San Francisco Community Broadband Opportunities Program
City of Chicago (Illinois)	SmartChicago Sustainable Broadband Adoption
Connect Arkansas (Arkansas)	Expanding Broadband Use in Arkansas Through Education
Connect Tennessee (Tennessee)	Computers 4 Kids: Preparing Tennessee's Next Generation forSuccess
District of Columbia Government (District of Columbia)	DC Broadband Education, Training and Adoption
Eastern Upper Peninsula Intermediate School District (Michigan)	Sparking Broadband Use in the Upper Peninsula of Michigan

Table 3. (Continued)

Grantee (state)	Project(s)
Foundation for California Community Colleges (California)	California Connects
Future Generations Graduate School (West Virginia)	Equipping West Virginia's Fire and Rescue Squads with Technology and Training to Serve Communities
Mexican Institute of Greater Houston (Texas)	Sustainable Broadband Adoption through Training for HispanicAdults
Michigan State University (Michigan)	Broadband Adoption through Education and E-Entrepreneurship inMichigan's Urban Cores
Tampa Housing Authority (Florida)	AccessALL Tampa
The Urban Affairs Coalition (Pennsylvania)	Freedom Rings: Sustainable Broadband Adoption

Source: GAO review of NTIA documents. | GAO-15-473.

To determine the extent to which FCC and NTIA set goals and assessed their performance to address broadband adoption barriers, we reviewed the agencies' most recent strategic plans and performance plans to identify their goals and performance measures related to broadband adoption. We reviewed documentation, including the design and final report, from NTIA's program evaluation of BTOP. We also reviewed available third-party program evaluations done of the 15 BTOP projects listed previously in table 3.[4] We interviewed FCC and NTIA officials about their efforts to evaluate the effectiveness of their broadband adoption efforts and about their strategic goals and performance measures for these efforts. We evaluated their efforts against our criteria for evaluating agency performance as established in our prior work and the GPRA Modernization Act of 2010.[5]

We conducted this performance audit from August 2014 to June 2015 in accordance with generally accepted government auditing standards. Those standards require that we plan and perform the audit to obtain sufficient, appropriate evidence to provide a reasonable basis for our findings and conclusions based on our audit objectives. We believe that the evidence obtained provides a reasonable basis for our findings and conclusions based on our audit objectives.

Appendix II: Data on Household Broadband Adoption Overall and by Demographic Groups, 2009 and 2013

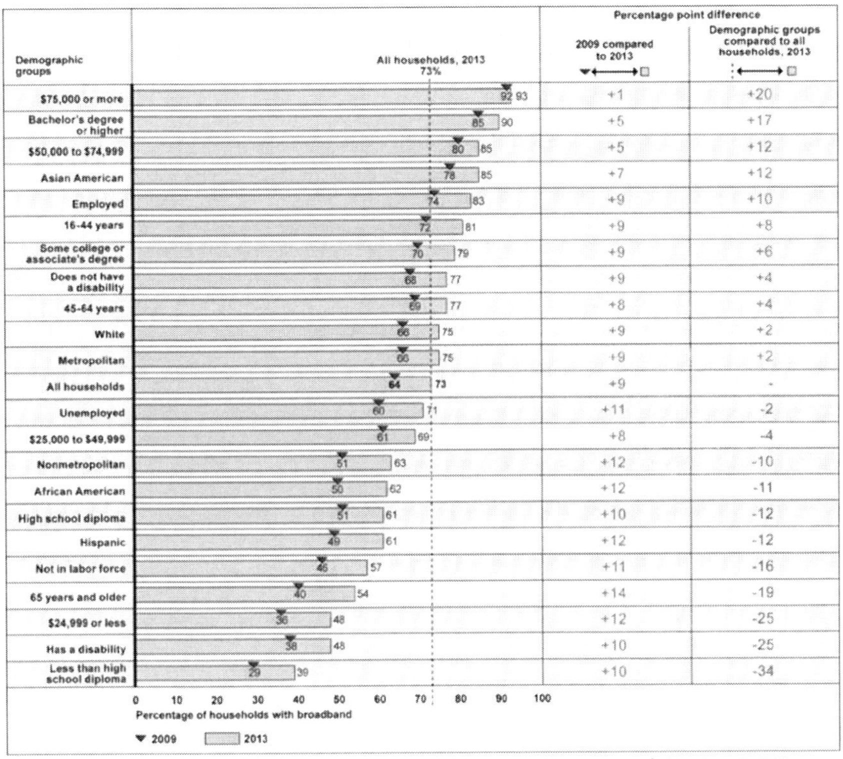

Source: US Census Current Population Survey and GAO amalysis. | GAO-15-473.

Note: The Census Current Population Survey asks the respondent if anyone in the household uses the Internet from home and then asks through which types of service the Internet is accessed. Respondents who say they use dial-up are not counted as having broadband. Respondents who indicate they access the Internet through DSL, cable modem, fiber-optic service, mobile broadband plan, satellite, or some other service are counted as a household with broadband. The household broadband adoption percentages we present are estimates calculated with intervals of 95 percent confidence. They have a sampling error of no larger than plus or minus 2.6 percentage points at the 95 percent level of confidence. Where we've drawn comparisons between groups or years to show differences, those differences are statistically significant and the confidence intervals do not overlap, except in the 2013 comparison between all households (95 percent confidence interval of 72.70 to 73.69 percent) and unemployed (95 percent confidence interval of 68.36 to 73.58 percent). About 1.6 percent of household respondents indicated they were more than one race; we have included these householders under multiple demographic race categories. Starting in 2010, Census changed the methodology for its income variable, including imputing household values that had been missing in prior years to provide values without missing observations.

Figure 2. Household Broadband Adoption among Demographic Groups Sorted by Adoption Percentage, 2009 and 2013.

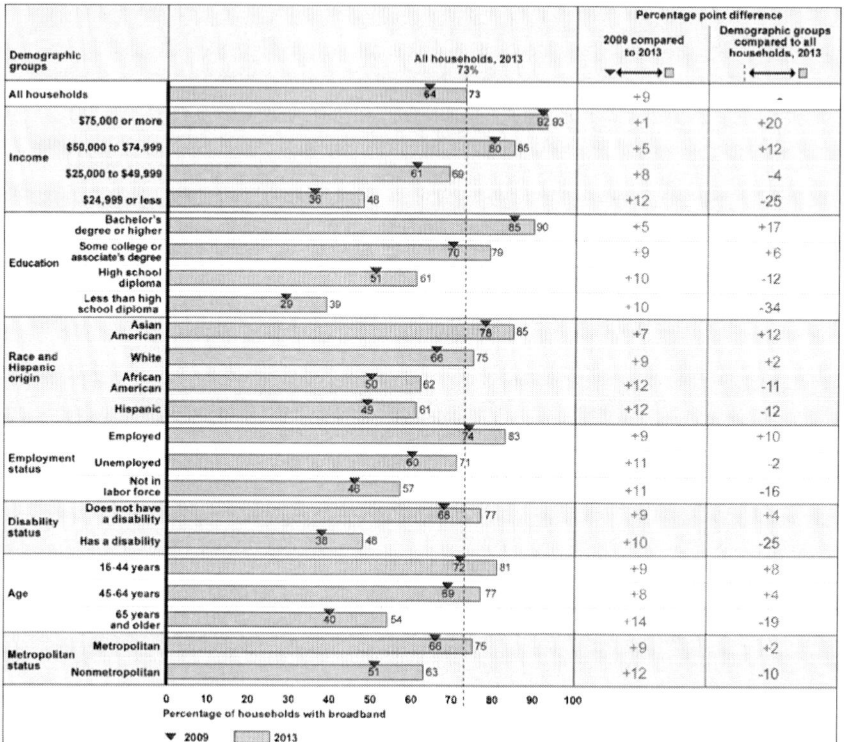

Source: US Census Current Population Survey and GAO amalysis. | GAO-15-473.

Note: The Census Current Population Survey asks the respondent if anyone in the household uses the Internet from home and then asks through which types of service the Internet is accessed. Respondents who say they use dial-up are not counted as having broadband. Respondents who indicate they access the Internet through DSL, cable modem, fiber-optic service, mobile broadband plan, satellite, or some other service are counted as a household with broadband. The household broadband adoption percentages we present are estimates calculated with intervals of 95 percent confidence. They have a sampling error of no larger than plus or minus 2.6 percentage points at the 95 percent level of confidence. Where we've drawn comparisons between groups or years to show differences, those differences are statistically significant and the confidence intervals do not overlap, except in the 2013 comparison between all households (95 percent confidence interval of 72.70 to 73.69 percent) and unemployed (95 percent confidence interval of 68.36 to 73.58 percent). About 1.6 percent of household respondents indicated they were more than one race; we have included these householders under multiple demographic race categories. Starting in 2010, Census changed the methodology for its income variable, including imputing household values that had been missing in prior years to provide values without missing observations.

End Notes

1 FCC, Inquiry Concerning the Deployment of Advanced Telecommunications Capability to All Americans in a Reasonable and Timely Fashion, and Possible Steps to Accelerate Such Deployment Pursuant to Section 706 of the Telecommunications Act of 1996, as Amended by the Broadband Data Improvement Act, GN Docket No. 14-126, 2015 Broadband Progress Report and Notice of Inquiry on Immediate Action to Accelerate Deployment, FCC 15-10 (rel. Feb. 4, 2015) (2015 Broadband Progress Report). The broadband availability data FCC presented in this report were collected by the National Telecommunications and Information Administration and the states in coordination with FCC. These data apply to types of fixed broadband, which are services that provide high-speed Internet access to a fixed point, such as cable, fiber, and digital subscriber line (DSL). The data exclude mobile and satellite services. FCC defines broadband, referred to in statute as "advanced telecommunications capability," based on capacity or speed. 47 U.S.C. § 1302(d)(1).The term "broadband" commonly refers to Internet service with speeds generally higher than dial-up connections, and our prior work has found that consumers may not be aware of their Internet speed. See GAO, Broadband Performance: Additional Actions Could Help FCC Evaluate its Efforts to Inform Consumers, GAO-15-363 (Washington, D.C.: Apr. 17, 2015). For these reasons, unless we are discussing FCC's broadband availability data, we are using "broadband" more generically without reference to a specific speed.

2 Pub. L. No. 111-5, § 6001(k)(1), 123 Stat. 115, 515-516 (2009).

3 FCC, Connecting America: The National Broadband Plan (Mar. 16, 2010).

4 In addition to increasing broadband adoption, BTOP also aimed to deploy broadband to underserved areas, expand the capacity of public computer centers, and develop and maintain a map showing the availability of broadband.

5 We limited the scope of literature to focus on (1) home broadband, as opposed to mobile broadband, (2) experience within the United States, and (3) articles from recent years just before and since the publication of FCC's National Broadband Plan in 2010.

6 "Technology organizations" refers to organizations that seek to help consumers get online through partnerships with broadband providers and community organizations.

7 Our review was not comprehensive of all efforts being taken to address broadband adoption barriers, since our purpose was to identify types of approaches used, not to catalog all efforts. We examined both federal efforts by NTIA and FCC and nonfederal efforts by the private sector and nonprofit organizations.

[8] We selected projects that met their goals because we were interested in identifying the most effective approaches to addressing broadband adoption barriers and reasoned that projects that met their goals were more likely to have been successful at this.

[9] NTIA did not require BTOP projects to have program evaluations; however, some projects elected to do so. We determined which projects had third-party program evaluations and obtained copies of the resulting reports from either NTIA or the project grantees. We reviewed these evaluations to determine which ones specifically examined the effectiveness of projects' approaches for addressing broadband adoption barriers.

[10] One grantee we interviewed administered two of the BTOP projects we examined. Additionally, another grantee we did not interview also administered two projects, resulting in 42 grantees for the 44 projects.

[11] GAO, Designing Evaluations: 2012 Revision, GAO-12-208G (Washington, D.C.: January 2012). The GPRA Modernization Act of 2010, Pub. L. No. 111-352, 124 Stat. 3866 (2011), amends provisions of the Government Performance and Results Act of 1993 (GPRA), Pub. L. No. 103-62, 107 Stat. 285 (1993).

[12] We did not assess FCC's Lifeline broadband pilot as part of this report. Rather, FCC's Lifeline program, including its broadband pilot, was the subject of our March 2015 report in which we examined the status of FCC's efforts to reform the program, the extent to which households participate in the program and face challenges in accessing and retaining its benefits, and FCC's plans to evaluate the broadband pilot. GAO, Telecommunications: FCC Should Evaluate the Efficiency and Effectiveness of the Lifeline Program, GAO-15-335 (Washington, D.C.: Mar. 24, 2015).

[13] Although NTIA initially funded 44 projects, the City of Tallahassee terminated its grant, leaving 43 projects that were completed or ongoing as of November 2014. NTIA reported in June 2012 that the City of Tallahassee had returned its $1.2 million award.

[14] National Broadband Plan.

[15] For example, the plan recommended that FCC provide free or low-cost broadband for low-income Americans, that NTIA form partnerships with public and private entities to promote broadband benefits, and that both agencies work together to create a national program to help Americans learn to use computers and the Internet.

[16] The 95 percent confidence interval for the 2013 estimate is 72.7 to 73.7 percent, and the associated 95 percent confidence interval for the 2009 estimate is 63.5 to 64.4 percent. These estimates are from the Computer and Internet Use supplement sponsored by NTIA, a supplement to the Current Population Survey, a national survey administered by the U.S. Census Bureau. For the purposes of the survey, respondents are identified as having broadband if they respond that their Internet service is provided by means other than a dial-up connection, such as cable or fiber. This is in contrast to data from FCC on the availability of advanced telecommunications capability, which are based on upload and download speed data submitted by service providers. The estimates derived from Current Population Survey data that are presented in this section are based upon the responses of the household member who answered the survey.

[17] The 95 percent confidence interval associated with the 2009 estimate is 39.1 to 41.1 percent, and the 95 percent confidence interval associated with the 2013 estimate is 53.2 to 55.4 percent.

[18] The 95 percent confidence interval associated with the 2009 estimate is 47.0 to 50.0 percent, and the 95 percent confidence interval associated with the 2013 estimate is 59.7 to 63.0 percent.

[19] The 95 percent confidence interval for the 2013 nonmetropolitan rate is 61.5 to 64.0 percent, and for the 2013 metropolitan rate it is 74.6 to 75.7 percent.

[20] According to the U.S. Census, the state-level adoption rates have margins of error no greater than plus or minus 0.8 percentage points at the 90 percent confidence interval, and the national average has a margin of error of plus or minus 0.1 percentage points at the 90 percent level of confidence.

[21] By "home" broadband, we are referring to broadband service to a home that residents can use with desktop or laptop computers, which is generally provided by wired means such as cable or fiber. This is in contrast to mobile broadband that is accessed by mobile devices (smartphones, tablets) through wireless means. Our scope in discussing broadband benefits is concerned primarily with those related to adoption at home.

[22] FCC, FCC and Connect to Compete Tackle Broadband Adoption Challenge, Oct. 13, 2011.

[23] ASR Analytics, Study Design Deliverables A-B: Broadband Technology Opportunities Program Evaluation Study (Potomac, Md.: Revised Jan. 30, 2012).

[24] Telemedicine includes telehealth, which is the use of medical information exchanged from one site to another via electronic communications (such as video or email) to improve a patient's clinical health status. Telehealth can include telemental health, which is the provision of mental health services to patients living in remote locations or otherwise underserved area.

[25] GAO, Telecommunications: FCC's Performance Management Weaknesses Could Jeopardize Proposed Reforms of the Rural Health Care Program, GAO-11-27 (Washington, D.C.: Nov.17, 2010).

[26] GAO, Telecommunications: USDA Should Evaluate the Performance of the Rural Broadband Loan Program, GAO-14-471 (Washington, D.C.: May 22, 2014), and GAO, U.S. Postal Service: Urgent Action Needed to Achieve Financial Sustainability, GAO-13-562T (Washington, D.C.: Apr. 17, 2013).

[27] Pew Research Center, Key Indicators in Media & News, State of the News Media 2014 (Mar. 26, 2014), accessed April 24, 2015, http://www.journalism.org/2014/03/26/state-ofthe-news-media-2014-key-indicators-in-media-and-news/.

[28] National Broadband Plan. Additionally, the plan identified accessibility for people with disabilities as a cross-cutting issue.

[29] Some stakeholders told us that broadband subscription cost data are difficult to obtain for reasons such as reluctance from broadband providers to share such information and carriers' bundling practices that make it difficult to separate broadband costs from the costs of other bundled services, such as telephone and video services. In 2011, FCC sought comment on whether it should collect price data from broadband providers, but this issue remains open for consideration. See In the Matter of Modernizing the FCC Form 477 Data Program, Report and Order, ¶ 13, 28 FCC Rcd. 9887 (2013).

[30] NTIA and Economics and Statistics Administration, Exploring the Digital Nation: America's Emerging Online Experience (June 2013). Using data from the 2011 U.S. Census Current Population Survey, NTIA reported estimates that 48 percent of households cited "don't need it" or "not interested" as the main reason for not using the Internet at home, followed by 28 percent that said "too expensive" and 13 percent that said they had "no computer" or "no adequate computer."

[31] See 2015 Broadband Progress Report, at Table 7. FCC is required by statute to determine whether advanced telecommunications capability is being deployed to all Americans in a reasonable and timely fashion, which is based, in part, on capacity or speed. 47 U.S.C. § 1302(b), (d)(1). In 2015, FCC increased the minimum speed benchmark to download speeds

of at least 25 megabits per second (Mbps) and upload speeds of at least 3 Mbps. Prior to this change, the minimum speed benchmark was at least 4 Mbps for downloads and 1 Mbps for uploads (although FCC's data used proxies of 3 Mbps for downloads and 768 kilobits per seconds for uploads). See 2015 Broadband Progress Report at para. 3. We're presenting a range of numbers to reflect broadband availability at both of these speed benchmarks. As previously stated, consumers are often not aware of their Internet speed. See Inquiry Concerning the Deployment of Advanced Telecommunications Capability to All Americans in a Reasonable and Timely Fashion, and Possible Steps to Accelerate Such Deployment Pursuant to Section 706 of the Telecommunications Act of 1996, as Amended by the Broadband Data Improvement Act, 27 FCC Rcd. 10343 (2012) (Eighth Broadband Progress Report).

[32] Our review is meant to illustrate these various approaches and is not comprehensive of all efforts to address broadband adoption barriers in the United States.

[33] According to NTIA's November 2014 quarterly report to Congress on BTOP, recipients of BTOP grants—both the Sustainable Broadband Adoption grant and the Public Computing Centers grant—had spent 99 percent of their available grant funding as of June 30, 2014. The report also states that 40 of the 44 broadband adoption grantees had closed out their projects with the remainder nearing completion. NTIA, Broadband Technology Opportunities Program (BTOP) Quarterly Program Status Report (November 2014).

[34] GAO-15-335.

[35] We recently initiated work examining access to telecommunications services, including broadband, on tribal lands; we expect to have the results of this work available in December 2015. Therefore, we did not include broadband issues on tribal lands within the scope of this report.

[36] The Recovery Act required BTOP funding recipients to report quarterly on their use of funds and NTIA to make these reports available to the public. See Pub. L. No. 111-5, div. B, tit. VI, § 6001(i)(1) (2009). NTIA also required that funding recipients report quarterly on their broadband equipment purchases and progress made in achieving goals, objectives, and milestones identified in the recipient's application. In addition, BTOP broadband adoption grantees had to report a variety of project-specific information, including the number of new household broadband subscriptions as a result of the program. 74 Fed. Reg. 33104, 33125 (July 9, 2009).

[37] A program evaluation is a systematic study using research methods to collect and analyze data to assess how well a program is working and why. Evaluations answer specific questions about program performance and may focus on assessing program operations or results. Evaluation results may be used to assess a program's effectiveness, identify how to improve performance, or guide resource allocation. See GAO-12-208G.

[38] NTIA, Broadband Adoption Toolkit (May 2013), accessed February 10, 2015, http://www2.ntia.doc.gov/files/toolkit_042913.pdf.

[39] ASR Analytics, Final Report: Social and Economic Impacts of the Broadband Technology Opportunities Program (Potomac, Md.: Submitted Sept. 15, 2014), accessed February 11, 2015, http://www2.ntia.doc.gov/files/asr_final_report.pdf.

[40] Additionally, one project did its own program evaluation. Some of the evaluation work associated with the BTOP projects was focused on outcomes other than broadband adoption, such as improved computer literacy and job prospects.

[41] GAO-15-335. We did not recommend changes to the broadband pilot to address this issue because the pilot's projects were substantially complete. Our report did recommend that FCC conduct a program evaluation to determine the extent to which the Lifeline program is

reaching its performance goals. In comments on the report, FCC agreed that it should evaluate the extent to which the Lifeline program is efficiently and effectively reaching its performance goals and said that FCC staff will address GAO's recommendation.

[42] GAO-12-208G.

[43] GAO, Program Evaluation: Studies Helped Agencies Measure or Explain Program Performance, GAO/GGD-00-204 (Washington, D.C.: Sept. 29, 2000) and Program Evaluation: Strategies to Facilitate Agencies' Use of Evaluation in Program Management and Policy Making, GAO-13-570 (Washington, D.C.: June 26, 2013).

[44] BTOP broadband adoption grant applicants stated in their grant applications how many new broadband household or business subscribers they estimated would result from their efforts. NTIA officials told us they consulted with projects that received grants and, in some cases, adjusted these goals. The projects we state did not meet their household subscriber goals are those that met neither their initial nor adjusted goals for households. Of the 43 completed projects, 15 met their adjusted goals, including 5 that reported twice as many or more subscribers than their adjusted goal. Of the 27 projects that did not meet their goals, 8 of the projects met at least 75 percent of their goal.

[45] "About NTIA" at http://www.ntia.doc.gov/about (accessed Feb. 13, 2015).

[46] The target for 2015 is 175 communities, and the target for 2016 is 250 communities.

[47] NTIA's fiscal year 2016 budget request to Congress includes $13 million for broadband programs described as continuing the momentum of BTOP by providing technical assistance and outreach to communities and stakeholder groups to address issues including broadband adoption. This funding represents a significant reduction as compared to the about $25 million NTIA spent on broadband programs for fiscal year 2013 when BTOP was fully active.

[48] GAO, Managing for Results: Executive Branch Should More Fully Implement the GPRA Modernization Act to Address Pressing Governance Challenges, GAO-13-518 (Washington, D.C.: June 26, 2013).

End Notes for Appendix I

[1] These estimates come from the Computer and Internet Use Supplement of the Current Population Survey, a national household survey sponsored by NTIA and administered by the U.S. Census Bureau. For purposes of this survey, respondents are identified as having broadband if they respond that their Internet service is provided by means other than a dial-up connection, such as cable or fiber. The estimates derived from these survey data are based upon the responses of the household reference person—i.e., the person that answered the survey.

[2] Our review was not comprehensive of all efforts being taken to address broadband adoption barriers, since our purpose was to identify types of approaches used, not to catalog all efforts.

[3] NTIA did not require BTOP projects to have program evaluations; however, some projects elected to do so. We determined which projects had third-party program evaluations and obtained copies of the resulting reports from either NTIA or the project grantees or through our own research. We reviewed these evaluations to determine which ones specifically examined the effectiveness of projects' approaches for addressing broadband adoption barriers.

[4] Since NTIA did not require BTOP grantees to have program evaluations of their BTOP projects, not every project we examined had an associated evaluation. We reviewed evaluation reports that we obtained from the grantees, NTIA, and our research.

[5] The GPRA Modernization Act of 2010, Pub. L. No. 111-352, 124 Stat. 3866 (2011), amends provisions of the Government Performance and Results Act of 1993 (GPRA), Pub. L. No. 103-62, 107 Stat. 285 (1993).

In: U.S. Broadband Issues
Editor: Chester Rios
ISBN: 978-1-63483-944-0

Chapter 2

BROADBAND PERFORMANCE: ADDITIONAL ACTIONS COULD HELP FCC EVALUATE ITS EFFORTS TO INFORM CONSUMERS*

United States Government Accountability Office

ABBREVIATION

ACSI	American Customer Satisfaction Index
CFPB	Consumer Financial Protection Bureau
DSL	digital subscriber line
FCC	Federal Communications Commission
ISP	Internet service provider
MBA	Measuring Broadband America
Mbps	megabits per second
NPRM	Notice of Proposed Rulemaking
UK	United Kingdom

* This is an edited, reformatted and augmented version of The United States Government Accountability Office publication, GAO-15-363, dated April 2015.

WHY GAO DID THIS STUDY

Broadband is increasingly seen as an essential communications service, with applications in education, medicine, public safety, and entertainment. However, consumers can experience broadband performance problems. FCC, which has primary responsibility for regulating broadband, has taken steps to measure broadband performance and to require that ISPs give consumers information about the performance of their services.

GAO was asked to review issues related to broadband performance information. This report examines (1) broadband performance information available to consumers and its limitations, if any, and (2) FCC's actions to evaluate its efforts to provide consumers with broadband performance information.

To address these objectives, GAO reviewed FCC proceedings; conducted a literature review; analyzed comments filed with FCC regarding broadband performance information; and interviewed FCC officials and various stakeholders from industry and public interest groups.

WHAT GAO RECOMMENDS

FCC should take additional steps to evaluate its efforts to provide consumers with broadband performance information. This should include: (1) conducting or commissioning research on the effectiveness of its efforts and making the results publicly available, and (2) establishing performance goals and measures that allow FCC to monitor and report on these efforts. FCC concurred with GAO's recommendations.

WHAT GAO FOUND

Consumers can access broadband performance information from several sources, including Internet service providers (ISP), online speed tests, and the Federal Communications Commission (FCC); however, the information has some limitations. For example:

- *ISP information:* FCC's transparency rule requires ISPs to disclose broadband performance information, but ISPs' disclosures vary, and some stakeholders said that the lack of standardization of disclosures can make it difficult for consumers to compare broadband services. Some ISPs, however, question the need and benefit of standardized disclosures. FCC recently adopted enhancements to the transparency rule, such as specifying what information ISPs must disclose, and FCC is considering potential disclosure formats.

- *Speed tests:* Consumers can use information from these tests to verify their broadband speeds. However, speed tests can be affected by many factors and may not detect congestion affecting a specific website. Thus, it can be difficult for consumers to identify the cause of their broadband performance problems.
- *FCC reports:* Through the Measuring Broadband America (MBA) program, FCC tests ISPs' networks and compares their actual and advertised speeds in an annual report. However, the report is not targeted toward consumers, and stakeholders stated that consumers may not be aware of the report.

FCC has taken steps to evaluate its efforts to provide consumers with broadband performance information, such as its transparency rule and MBA program; however, FCC's ability to evaluate its efforts is limited by a lack of useful performance information and relevant performance goals and measures. For instance, while FCC obtains information about the effectiveness of its efforts from stakeholders' comments and consumers' complaints, FCC has not sought performance information from more objective sources, such as consumer research—as has been done by other government entities. Further, although consumer complaints can provide FCC with valuable insights on topics of interest to the commission, FCC has acknowledged that complaints may not provide a complete picture of consumer's information needs and some industry stakeholders have questioned FCC's reliance on complaints as a basis for making decisions. In addition, although FCC's strategic plan includes strategic objectives related to informing consumers about broadband networks, FCC lacks performance goals and measures to monitor the impact and effectiveness of its efforts to provide consumers with broadband performance information. GAO has previously reported that critical elements of effective performance management include information that is complete and consistent, with performance measures linked to goals. Without such information and measures, FCC cannot be assured that its efforts to provide consumers with broadband performance information are effective and meeting consumers' needs.

April 17, 2015

The Honorable Frank Pallone, Jr.
Ranking Member
Committee on Energy and Commerce House of Representatives

The Honorable Anna G. Eshoo
Ranking Member
Subcommittee on Communications and Technology
Committee on Energy and Commerce

House of Representatives

The Honorable Edward J. Markey
United States Senate

Broadband is increasingly seen as an essential communications service, with applications in education, medicine, public safety, and entertainment. Broadband allows people to share large amounts of information at increasingly faster speeds and provides access to information otherwise unavailable, such as online courses not offered in their local area. Press reports indicate that consumers spent over $100 billion on Internet access in 2013, and the Federal Communications Commission (FCC) has noted that the number of fixed and mobile residential broadband connections in the United States grew from 48 million in December 2010 to nearly 180 million in December 2013.[1] Consumers' use of broadband has changed over the years as well, with more people using broadband to access video and data-intensive applications, and this trend is expected to continue. For example, Cisco (a provider of network infrastructure) projects that consumer Internet traffic in North America will grow at an average annual rate of 21.5 percent between 2013 and 2018.[2] However, FCC has previously expressed concerns regarding consumers' access to accurate and useful information about their broadband service performance. In addition, GAO has previously identified limitations in the broadband performance information available to consumers, and recently noted the need to ensure consumers understand issues that can affect their use of broadband, such as usage-based pricing.[3]

You asked us to review issues related to broadband performance information available to consumers. This report examines: (1) the information currently available to consumers regarding broadband performance and the limitations of this information, if any, and (2) FCC's actions to evaluate its efforts to provide consumers with broadband performance information. To identify the information currently available to consumers regarding broadband performance and any limitations of the information, we reviewed prior GAO reports on this issue, as well as relevant FCC proceedings, reports, and documents—including its Measuring Broadband America reports, 2010 *Open Internet* Order,[4] and 2014 *Open Internet* Notice of Proposed Rulemaking (NPRM) and subsequent 2015 *Open Internet* Order.[5] We also reviewed broadband provider websites, third-party sites that provide broadband performance information to consumers, and studies on broadband performance measurement. To identify limitations, we reviewed comments filed with FCC regarding the effectiveness of its transparency rule and Measuring Broadband America effort. To select relevant comments, we performed key-word searches that included terms related to transparency and disclosure for consumers. To ensure we reviewed a broad range of comments, we selected comments from stakeholders that represented a variety of interests, including public interest groups, Internet service providers (ISP), content providers, trade associations, and consumer representatives. We also interviewed representatives from FCC and selected stakeholder groups, including public interest and consumer advocacy groups, industry associations, content or equipment providers, ISPs, and other industry stakeholders. We identified stakeholders to interview based on our review of comments filed in FCC's 2014 *Open Internet* and related proceedings, as well as based on recommendations from other organizations we interviewed.

To determine how FCC evaluates its efforts to provide consumers with broadband performance information, we reviewed relevant FCC documents, such as FCC's strategic and performance plans. We also reviewed FCC's requests for comment on consumers' information needs and the effectiveness of FCC's efforts to provide consumers with broadband performance information, as well as selected comments filed in response to these proceedings (as described above). We also reviewed prior GAO reports on FCC and results-oriented management and compared FCC's use of performance information and measures to performance management best practices outlined in previous GAO work. To further inform our assessment, we interviewed representatives from the stakeholder groups identified above, as well as officials from the Federal Trade Commission and National Telecommunications and Information Administration regarding their efforts to inform consumers about broadband issues and to evaluate the effectiveness of those efforts. In addition, we reviewed reports and studies on consumer disclosures conducted by FCC's Transparency Working Group; foreign telecommunications regulatory bodies, including those in the United Kingdom and European Union; and other federal agencies, such as the Consumer Financial Protection Bureau. See appendix I for more information about our scope and methodology.

We conducted this performance audit from July 2014 to April 2015 in accordance with generally accepted government auditing standards. Those standards require that we plan and perform the audit to obtain sufficient, appropriate evidence to provide a reasonable basis for our findings and conclusions based on our audit objectives. We believe that the evidence obtained provides a reasonable basis for our findings and conclusions based on our audit objectives.

BACKGROUND

The term "broadband" commonly refers to Internet access that is high speed and provides an "always-on" connection, so users do not have to reestablish a connection each time they access the Internet. Broadband speeds are described in download and upload capabilities, and are defined as the maximum rates at which data are delivered over the Internet. A higher speed indicates a faster information delivery rate. For example, a 10 megabits per second (Mbps) service should deliver ten times as much data as a 1 Mbps service in the same period of time.

Internet service providers (ISP) have marketed and competed with one another on the speed of their various broadband plans. Advances in technology, such as the use of fiber (described below) and new wireless technologies, have allowed ISPs to offer increasingly faster speeds that support new services and applications, such as streaming video. FCC takes those advances into account as part of its required inquiry into whether "advanced telecommunications capability" is being deployed to all Americans in a reasonable and timely fashion.[6] Pursuant to this inquiry, FCC recently updated the minimum speed benchmark it uses to measure the availability of advanced telecommunications capability.[7] In addition to speed, a number of factors can be used to evaluate broadband performance. See table 1.

Table 1. Measures of Broadband Performance

Term	Description	How measure relates to the end-user's experience
Upload Speed	Upload speed refers to the rate at which data are transferred from the consumer to the Internet. Measured by the number of bits of data transferred per second.	Slow upload speeds can create delays when consumers aretrying to upload content to the Internet, such as uploading photos to a website or using video chat.
Download Speed	Download speed refers to the rate at which data are transferred to the consumer. Measured by the number of bits of data transferred per second.	Slow download speeds can create delays when consumers are accessing Internet content, which can manifest as long waits for the loading of webpages and problems in viewing video content.
Latency	A measure of the time that it takes for data to travel from a computer to a server and back again.	Accessing Internet content requires multiple messages to pass between the end-user and the web server, so latency can have a significant impact on broadband performance. A high-latency network connection experiences long delay times. High latency can affect the performance of online gaming, videoconferencing, phone, and streaming media services.
Jitter	Jitter measures the variance in latency. Zero jitter means the results were exactly the same every time, and anything above zero is the amount by which they varied. A lower jitter value is better.	High jitter can affect consumers' ability to use real-time applications such as video or audio applications.
Packet Loss	A measure of the amount of data lost in transmission from one device to another.	Packet loss can result in slower download and upload speeds, poor quality Internet phone services, and pauseswith streaming media.
Availability	Refers to the amount of time that the network is functioning, stable and able to provide connectivity between a computer and a desired application.	If a network experiences an outage, a consumer will be unable to use their broadband service.

Source: GAO analysis of FCC, academic, private sector, and other federal sources. | GAO-15-363.

Note: This list is not comprehensive, as there are many technical measures that may be used to evaluate specific aspects of broadband performance.

Consumers generally subscribe to broadband in two ways:

- Fixed: In-home fixed Internet plans are often sold as a monthly subscription by cable television or telephone companies. Service from cable television companies is generally provided through the same coaxial cables that deliver television programming. Service from telephone companies is generally provided through traditional copper telephone lines—commonly referred to as digital subscriber line (DSL) service—or fiber-optic lines, which convert electrical signals carrying data into light and send the light through glass fibers. These network technologies generally have higher speeds than mobile networks. Consumers can connect a variety of devices to in-home fixed networks through a wired connection or wireless Wi-Fi connection (see fig. 1). Consumers are increasingly using the Internet to supplement or replace their use of traditional services, such as traditional telephone and cable TV service.
- Mobile: Traditionally, mobile providers sold access to the Internet as an add-on to mobile telephone service plans that may or may not include a multiyear contract. Mobile service is provided through cell tower coverage with data sharing through radio spectrum.[8] A number of devices may connect to mobile broadband networks, such as smart phones, tablets, and mobile devices that enable laptops to connect to mobile networks (see fig. 2).

The 2014 American Customer Satisfaction Index (ACSI)[9] reported that with respect to fixed broadband service, consumer satisfaction regarding data speeds and video streaming increased from 2013 to 2014. However, ACSI also found that during this time period, customer satisfaction related to service interruptions and outages, speed and service reliability, and peak evening-hour performance declined. With respect to mobile broadband providers, ACSI reported that customer satisfaction related to speed for downloading data and streaming content increased from 2013 to 2014; however, ACSI also reported that the customer satisfaction scores indicated that improvements were still needed in speed and reliability.

Broadband performance problems can manifest in a number of ways, such as slow loading of content, lost connections, and pauses in streaming video and audio.

Source: GAO | GAO-15-363.

Figure 1. Examples of Devices Connected to In-Home Fixed Broadband Service.

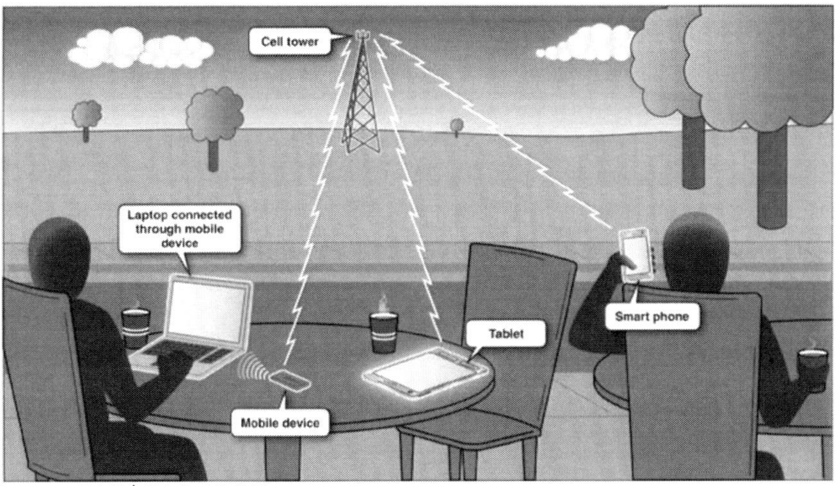

Source: GAO | GAO-15-363.

Figure 2. Examples of Devices Connected to Mobile Broadband Network.

FCC, ISPs, content providers, and researchers have noted that a variety of factors can affect broadband performance, such as accessing the Internet during peak usage times,[10] outdated equipment in the home, or multiple users in the same household accessing the Internet simultaneously. In addition, mobile broadband performance will vary based on the location of the phone or device.[11] Running multiple applications at once or having multiple users in a household may require greater speeds. Different types of Internet applications use varying amounts of data and require different minimum speeds to operate properly (see table 2). For example, email and web-browsing use small amounts of data, while watching movies uses larger amounts of data. Use of data-intensive broadband applications is projected to grow.[12] FCC has noted that consumers are increasingly using broadband applications that require large amounts of data, such as streaming video content, which can require more consistency of service and be more visibly affected by technical problems.

Table 2. Minimum Download Speeds for Different Broadband Activities (Assumes One Activity Is Occurring at a Time), as of February 2015

Activity	Minimum downloadspeed in megabits persecond (Mbps)
Email	0.5
Web browsing	
Job searching, navigating government websites	0.5
Interactive pages and short educational videos	1
Streaming radio	Less than 0.5
Phone calls/Voice over Internet Protocol[a]	Less than 0.5
Watching video	
Standard-definition streaming videos	0.7
Streaming feature movies	1.5
High-definition quality streaming movie or university lecture	5—8
Video conferencing	
Basic video conferencing	1
High-definition video conference and tele-learning	1- 2
Gaming	
Game console connecting to the Internet	1
Two-way online gaming in high-definition	4

Note: The entire citation for the source is: Federal Communications Commission's Broadband Speed Guide (http://www.fcc.gov/guides/broadband-speed-guide), last accessed February 10, 2015, and 2015 Broadband Progress Report.

[a] Voice over Internet Protocol is a technology that allows users to make voice calls using a broadband Internet connection instead of a regular phone line.

In addition, congestion at various points on the network can affect a user's broadband performance. In order to connect their customers to the Internet, ISPs link with other networks, including "backbone providers" that move Internet traffic long distances and interconnect with other networks. Companies may use a number of options to deliver their content across the Internet, such as connecting with a backbone provider (directly, or through an ISP), connecting directly to the consumer's ISP, or paying a content delivery network to deliver their content to ISPs.[13] See figure 3 for a simplified illustration of how content moves from websites and applications to consumers. Congestion occurring within the content provider's network, at points of interconnection between the networks, or within the ISP's network can all affect a consumer's broadband performance. ISPs' traffic management policies can also affect a consumer's broadband performance. This issue is discussed later in the report.

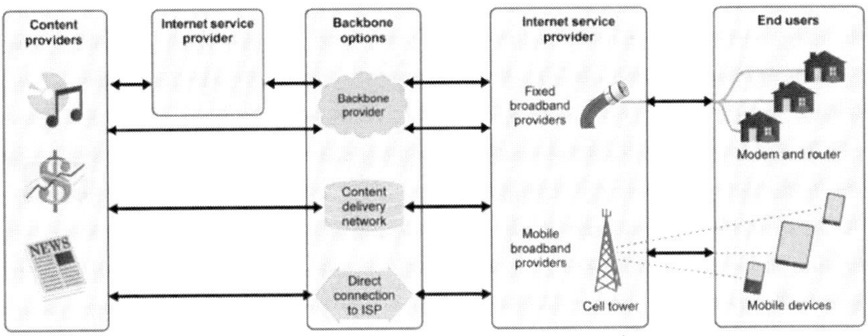

Source: GAO analysis of FCC industry and academic sources. | GAO-15-363.

Figure 3. Simplified Example of Internet Content Delivery.

FCC has primary responsibility for regulating broadband. In response to a requirement in the American Recovery and Reinvestment Act of 2009, FCC staff developed The National Broadband Plan, which includes a series of recommendations aimed at ensuring all Americans have access to broadband.[14] In addition, section 706 of the Telecommunications Act of 1996, as amended, directs FCC "to encourage the deployment on a reasonable and timely basis of advanced telecommunications capability to all Americans" through a variety of measures, including measures that promote competition in the local telecommunications market.[15] FCC has repeatedly noted the connection between informed consumers and increased competition, noting that the latter is an important part of promoting broadband deployment.[16]

FCC adopted an *Open Internet* Report and Order in 2010[17] outlining certain limits on how ISPs could treat network traffic. The 2010 *Open Internet* Order imposed three requirements on broadband providers: transparency, anti-discrimination, and anti-blocking. After facing a court challenge that affirmed the transparency rule but vacated and remanded the antidiscrimination and anti-blocking rules, FCC initiated a new rulemaking to address these issues,[18] which resulted in the adoption of the 2015 *Open Internet* Order.[19] This order, adopted in February 2015, addressed issues related to ISPs' disclosure of broadband performance information to consumers and other parties, as well as other practices that may affect consumers' broadband performance. The 2015 *Open Internet* Order also responded to concerns that ISPs may have the incentive and ability to act in ways that limit Internet openness (e.g., by blocking or discriminating against unaffiliated applications and services).

Although Consumers Can Access a Number of Sources of Information on Broadband Performance, the Information Is Limited

Consumers can access information on broadband performance from a number of sources, including their ISPs, third-party websites, and FCC reports. FCC adopted a transparency rule requiring ISPs to provide consumers with information on their services' performance characteristics. However, the information available to consumers is not standardized, and some stakeholders stated that this lack of standardization makes it difficult for consumers to compare broadband services. In addition, the information available through speed tests may not accurately reflect the end-user's actual experience. FCC has taken steps to measure and report on the extent to which ISPs are providing the speeds they advertise. However, the reports are not targeted toward consumers, and stakeholders stated that consumers might not be aware of the information.

Performance Information from ISPs and Third Parties Is Not Standardized and May Not Accurately Reflect the End-User's Actual Experience

Consumers interested in purchasing broadband service can go to providers' websites to get information on performance, including speed and latency, as well as information on availability and price. ISPs may also provide tools for selecting the correct package based on use and information on factors that affect broadband performance.[20] ISPs told us that consumers may also call and speak to a representative to determine the best package for their needs, and existing customers may call for help troubleshooting problems with their broadband service.

FCC's 2010 *Open Internet* Order included a transparency rule intended to provide consumers with more information on broadband performance. The rule requires that fixed and mobile providers "publicly disclose accurate information regarding the network management practices, performance, and commercial terms of its broadband Internet access services sufficient for consumers to make informed choices regarding use of such services..." [21] In response to concerns from ISPs regarding the burden associated with specific requirements, FCC did not require ISPs to disclose specific information in a standardized format. Instead, the 2010 *Open Internet* Order stated that FCC expected disclosures to include some or all of certain types of information, such as actual speed, to be timely and prominently disclosed in plain language accessible to consumers and to, at a minimum, be provided through a publicly available, easily accessible website. As part of its 2014 *Open Internet* NPRM, FCC proposed enhancing the transparency rule and requested comment on a variety of options, some of which are discussed later in this report. In the resulting February 2015 *Open Internet* Order, FCC adopted enhancements to the transparency rule, and stated that ISPs are required to disclose certain information under the transparency rule. Specifically, FCC now states that disclosures of network performance should include actual speed, latency, and packet loss, and disclosures should also include network management practices applied to specific users or user groups (for example, users in a particular service plan or geographic location). FCC also requires ISPs to develop a mechanism for directly notifying consumers if their individual use of a network will trigger a network practice that is likely to have a significant impact on the consumer's use of the service.[22]

Currently, ISPs' disclosures vary with respect to length, content, and where they are placed on ISPs' websites. In addition, according to public interest groups we spoke with, the complexity of this information and its lack of standardization across ISPs can make it difficult for consumers to find and use the information to compare broadband products and services. Some groups also raised concerns that consumers have difficulty understanding the information. However, some of the ISP and trade associations we spoke with questioned the rationale for more specific disclosure requirements, stating that consumers have access to the information they need or that flexibility is important, given the difficulties in explaining technical performance information to consumers. In addition, some ISPs and a trade association argued in written comments filed with the FCC that a standardized disclosure requiring specific information from ISPs could overlook features that may be of greater importance to consumers (such as nationwide wireless access) or confuse consumers by providing highly technical information. However, other ISPs were open to adopting a standardized label (as discussed below), with one ISP stating in written comments that it might help consumers make informed choices.

FCC's Open Internet Advisory Committee's Transparency Working Group studied how ISPs' present performance and pricing information and issued a report in which it recommended that FCC promote a voluntary labeling program to help consumers more easily compare and select broadband service offerings. The working group suggested the program be voluntary and result in a label that would include information on an ISP's upload and download speed, price, and usage restrictions. However, the report also noted a number of complexities associated with implementing a standardized label program, such as accounting for variability in broadband speeds, as well as for factors beyond the ISP's control, and the potential for consumers to purchase more expensive packages due to a lack of consumer education on what speeds they need. FCC officials told us that they have considered standardized labels, but developing a label that is easy to use, useful for consumers with differing broadband needs, and relevant in a changing environment is complicated. Researchers have noted that standardized performance information could help third parties develop tools and information that allow consumers to compare multiple broadband services. FCC requested comment on the working group's label proposal as part of its 2014 *Open Internet* NPRM. In its 2015 *Open Internet* Order, FCC established a voluntary safe harbor for the format and nature of the required disclosure to consumers. To take advantage of the safe harbor, a broadband provider must provide a consumer-focused, stand-alone disclosure. FCC, however, did not mandate the exact format of such disclosure at this time. Instead, it directed its Consumer Advisory Committee to develop and submit to the commission a voluntary disclosure format that ISPs could use to ensure compliance with the transparency rule no later than October 31, 2015.[23]

Online third-party speed tests allow consumers to test the performance of their broadband service by sending and receiving data between the consumer's device (such as a laptop or smart phone) and a test server. For example, www.speedtest.net measures users' broadband speed, and www.pingtest.net tests other parameters and provides an overall grade that includes information on which applications may be affected by the user's broadband performance characteristics. FCC also released an application that allows users to test the speed, latency, and packet loss of their mobile broadband.[24] In addition, ISPs may provide speed test tools to help consumers understand their broadband performance. In some cases, third parties have created sites that rank broadband providers based on the results of speed-test data. For example, OOKLA's www.netindex.com uses data from its speed test and consumer surveys to create country, state, and city-level rankings of fixed and mobile broadband providers in terms of upload and download speed, as well as other performance measures. This allows consumers to compare providers based on their performance in the consumer's local area. RootMetrics.com provides similar data on mobile broadband, including assessments of the mobile ISPs' speed and reliability.

However, the information from speed tests may not provide consumers with the information they need to understand what factors are affecting their broadband performance. Studies we reviewed and stakeholders we spoke with noted that speed tests can be affected by a variety of factors, which makes it difficult for consumers to identify what specific problem is affecting their broadband performance. For example, researchers have noted that online speed tests can be affected by the home equipment and the number of users accessing the Internet in the home; however, these tests may not enable consumers to discern that these factors are affecting their speeds.[25] In addition, speed tests can be limited in reflecting the end-user's actual experience because speed tests vary methodologically, and different tests can provide differing results, depending on the location of the speed test server. If a speed test server is located outside the consumer's ISP network, then the results could be affected by congestion occurring on networks outside of the ISP's control. For example, when using third-party speed tests, a consumer's traffic may travel over networks that are not controlled by their ISP, and congestion on those networks may have an effect on the speed test. On the other hand, speed tests hosted by an ISP may only test the performance of the ISP's network, and thus do not capture congestion that may be affecting a consumer's ability to access content outside of their ISP's network.

Stakeholders we interviewed noted that consumers may experience performance problems due to disputes between ISPs and content providers regarding interconnection (interconnection disputes) that lead to network congestion when consumers are accessing specific sites, and these problems may not be reflected in speed tests. Congestion at interconnection points has been identified as a problem contributing to poor broadband performance, particularly with respect to streaming video content. Such congestion may not be adequately captured by third-party or ISP speed tests, since it may be limited to a specific interconnection point not used by the speed test server. Some content providers, such as Google, have provided alerts to consumers when Internet congestion is affecting their broadband performance. However, information regarding interconnection agreements can be subject to non-disclosure agreements, and there have been several disputes regarding who is responsible for addressing interconnection congestion. In June 2014, FCC released a statement asserting that these interconnection disputes can affect consumers' broadband performance and that in the interest of providing consumers with more transparency, FCC was obtaining copies of interconnection agreements to further its understanding of interconnection disputes. Separately, in the 2015 *Open Internet* Order, FCC stated that it would hear interconnection disputes on a case-by-case basis.

ISPs' traffic management policies can also affect the results of speed tests, making comparisons across providers difficult. In August 2014, a public interest group announced it was filing a complaint against T-Mobile. According to the group, T-Mobile was slowing down the broadband speeds provided to mobile broadband subscribers that had reached their usage limits;[26] however, T-Mobile did not slow down traffic coming from speed test applications. Thus, if a T-Mobile customer whose broadband was being slowed down accessed a speed test to try and determine their broadband speed, their speed test would show a higher speed than the customer was actually receiving. The public interest group stated that this was a violation of the transparency rule, because consumers were not receiving accurate information regarding network performance. In November, FCC issued a press release stating that T-Mobile agreed to take several actions to ensure consumers were able to obtain accurate measures of their speed, including providing a link to a speed test that would reflect the speed reduction. There have also been press reports that backbone networks' traffic management policies have affected speed test results, and researchers have noted that ISPs can design their networks to provide better speed test results.

FCC Reports on Broadband Performance, but the Reports Are Not Targeted toward Consumers

Since 2011, FCC has published annual reports with broadband performance results from its Measuring Broadband America (MBA) effort.[27] The MBA program compares fixed ISPs' average delivered download and upload speeds against their advertised speeds (for a 24- hour period, as well as for the peak usage time of 7 PM to 11 PM on weeknights), and provides additional information on measurements of latency and website loading time. FCC is working to expand its MBA effort to include mobile broadband providers and, as part of that process, has released an app for consumers to download and test their mobile broadband speeds.[28]

MBA developed out of a recommendation in the 2010 National Broadband Plan to improve the availability of information for consumers about their broadband service. At the time, there were concerns that ISPs were not delivering their advertised speeds, and data were not available to determine whether this was the case. FCC officials stated that MBA allowed them to verify the accuracy of ISPs' advertising claims. FCC collaborated with industry and public interest stakeholders when designing the MBA program. The sample population is drawn from subscribers of 14 of the largest ISPs (which serve over 80 percent of the residential marketplace) and consists of about 6,000 volunteers. Unlike online speed tests that consumers may access from a variety of websites, MBA only tests elements under the direct or indirect control of a consumer's ISP, from the consumer's gateway—the modem or router used by the consumer to access the Internet—to a nearby major Internet gateway point. Thus, the results are not affected by congestion occurring outside of an ISP's network.

In 2014, FCC found that the ISPs included in the report were, on average, delivering 101 percent of advertised download speeds during the peak usage hours.[29] FCC, ISPs, and public interest group representatives stated that the transparency provided through MBA spurred competition among providers and led to improved speeds, which benefit consumers.[30] In addition, ISPs that participate in the MBA effort can provide consumers with their MBA results in order to comply with the transparency rule's disclosure requirements. Researchers have used raw data from MBA to conduct in-depth analyses of broadband performance. FCC officials also told us that it uses the information from the MBA report to inform its rulemakings on broadband, such as raising the broadband speed for providers to be eligible for the Connect America Fund, which supports telephone and broadband service for rural residents.[31]

FCC is also in the process of considering next steps for the program. As noted above, in order to provide comparable results across ISPs, the MBA initially only measured the ISP's network. Thus, delays incurred in the consumer's home or in segments of the Internet outside an ISP's network were excluded from the results. For example, in the 2014 MBA report, FCC noted that it excluded data from the report after discovering some testing servers (and thus, any measurements taken using these servers) were affected by congestion at interconnection points where ISPs exchange traffic with Internet backbone providers.[32] In the report, FCC acknowledged that consumers accessing services and content over the affected paths would likely see a significant degradation in their service, but FCC stated that it did not include the results from these servers in its report because it aims to provide an analysis of average network performance. Two public interest groups we interviewed raised concerns that the MBA results were not reflecting the experience of consumers whose broadband performance was negatively affected by interconnection disputes. In the 2014 MBA report, FCC noted that while MBA was initially focused on measuring broadband performance from the consumer to the end of the service provider's network, Internet services and applications rely on a complex and variable arrangement of interconnected networks to reach consumers. FCC stated that it is exploring options to leverage its MBA effort to gain more insights into the evolving performance of the Internet and provide better information to consumers. For example, FCC officials told us that it is testing an expansion of the MBA program to include measuring broadband performance when a consumer is using streaming video services.

Although the program initially was developed to provide consumers access to broadband performance information, many of the various stakeholders we interviewed, as well as FCC, generally agreed that most consumers are unaware of the MBA report and do not rely on it to obtain broadband performance information. MBA reports are highly technical and include detailed appendices and raw data. FCC and some of the ISPs and public interest groups we interviewed noted that information collected through the MBA program may reach consumers through secondary sources, such as ISPs' advertising and transparency rule disclosures, news reports, and tech blogs. In addition, as noted above, FCC and many of the ISPs and public interest groups stated that the program has spurred competition among providers and led to improved speeds. However, ISPs' offerings and performance can vary geographically, and in the *Open Internet* comments we reviewed, two stakeholders raised concerns that the MBA reports do not provide consumers with information about broadband performance at the local level. The National Broadband Plan recommended that FCC's broadband performance reporting effort should include "detailed information about the actual performance of the country's top broadband service providers in different geographic markets (e.g., by county, city or [Metropolitan Statistical Area])." In the 2014 MBA report, FCC stated that the data are not collected in a way that permits meaningful conclusions about broadband performance at the local level and that this limitation was a result of cost issues and the finite number of measurement devices that could be deployed over the course of the project. A report on the MBA effort that was commissioned by FCC and conducted by the North Carolina State University's Institute for Advanced Analytics primarily reviewed MBA's methodology, but also made some suggestions for making the data more consumer-friendly, including the use of a tool developed by the group to help consumers compare ISPs' performance in their states (when enough data were available to allow for statistically valid comparisons).

FCC Has Taken Steps to Evaluate Its Efforts to Inform Consumers, but Could Enhance Its Performance Information and Measures

FCC has taken steps to determine whether its efforts to provide consumers with broadband performance information—such as the transparency rule and MBA program—are effective and meeting consumers' needs. Specifically, FCC has reviewed consumer complaints and requested stakeholder comments on this issue in several recent rulemaking proceedings. However, FCC's ability to evaluate its efforts is limited by a lack of useful performance information. In addition, FCC established strategic objectives related to informing consumers in its strategic plan, but lacks performance goals and measures to monitor the effectiveness of its efforts to provide consumers with broadband performance information. Additional actions—such as conducting or commissioning objective consumer research and establishing performance goals and measures—could help FCC evaluate whether its efforts to provide consumers with broadband performance information are effective and meeting consumers' needs.

FCC Has Requested Stakeholder Comments in Several Recent Rulemakings and Reviewed Consumer Complaints

FCC has requested stakeholder comments in rulemaking proceedings and reviewed consumer complaints filed with the commission to determine whether its efforts to provide consumers with broadband performance information—such as its transparency rule and MBA program—are effective and meeting consumers' needs. According to FCC officials, FCC relies heavily on stakeholder comments to inform and explain decisions underlying the commission's orders; however, FCC may also request stakeholders' comments to develop a body of knowledge on issues of interest to the commission. In several recent rulemaking proceedings, FCC has requested comments from stakeholders on (1) consumers' broadband performance information needs, (2) best practices for presenting and displaying performance information to consumers, and (3) potential improvements to FCC's efforts to provide consumers with broadband performance information, among other things. For instance, FCC officials told us that stakeholder comments were used as part of FCC's determination to enhance its transparency rule. Table 3 summarizes some of the consumer-related issues and questions FCC has raised and requested stakeholder comments on in recent years.

Table 3. GAO Analysis of Selected Consumer-Related Issues and Questions Raised in Recent Federal Communications Commission's Rulemaking Proceedings

Date	Federal Communications Commission's (FCC) Rulemaking Action	Summary of selected consumer-related issues and questions raised
August 2009	*Notice of Inquiry*[a]	•What information do consumers need when selecting a service provider and plan, managing their service plan, and deciding whether and when to switch providers or plans? •What information do consumers need to assess

Date	Federal Communications Commission's (FCC) Rulemaking Action	Summary of selected consumer-related issues and questions raised
		service quality? •What are best practices related to formatting and displaying information to consumers and measuring the effectiveness of consumer disclosures? •What additional steps could FCC take to improve its consumer education efforts?
June 2010	*Public Notice* [b]	•How should FCC measure performance of mobile broadband? •How should FCC convey mobile-broadband performance information to consumers—are there best practices FCC should follow? •What are the existing sources of mobile-broadband performance information available to consumers and how they could be improved?
April 2011	*Public Notice*[c]	•What types of broadband performance information are most useful to consumers assessing service options? •What is the most effective way to present the information and ensure Internet service providers (ISP) are providing this information to consumers?
May 2014	*Notice of Proposed Rulemaking*[d]	•Should FCC revise the transparency rule to require ISPs to disclose specific network practices, performance characteristics (e.g., effective download speeds, upload speeds, latency, and packet loss), and/or terms and conditions of service to end users (e.g., data caps)? •Are there best practices for displaying and formatting ISP disclosures for end users, such as the use of a standardized label, as suggested by the Transparency Working Group? •Are there areas of the Measuring Broadband America (MBA) program thatcould be improved to provide more useful information to consumers?

Source: GAO Analysis of Selected Federal Communications Commission Rulemaking Proceedings. | GAO-15-363.

[a] In the Matter of Consumer Information and Disclosure, Notice of Inquiry, 24 FCC Rcd. 11380 (2009).

[b] Comment Sought on Measurement of Mobile Broadband Performance and Coverage, Public Notice, 25 FCC Rcd. 7069 (2010).

[c] Consumer and Government Affair Bureau Seeks Comment on "Need for Speed" Information for Consumers of Broadband Services, Public Notice, 26 FCC Rcd. 5847(2011). FCC noted that although the transparency rule adopted in its 2010 Open Internet Order requires broadband providers to disclose information regarding network management practices, performance, and commercial terms of broadband services, the Notice represented a step further in the commission's ongoing effort to ensure that consumers have access to the information they need to make informed decisions.

[d] Protecting and Promoting the Open Internet, Notice of Proposed Rulemaking, 29 FCC Rcd. 5561 (2014).

According to FCC officials, FCC also relies on consumer complaints filed with the commission to determine whether its efforts to provide consumers with broadband performance information are effective and meeting consumers' needs. For example, in the 2014 *Open Internet* NPRM, FCC stated that it had received and analyzed hundreds of consumer complaints related to ISPs' disclosures under the transparency rule and that these complaints were part of the agency's rationale for tentatively concluding the rule should be strengthened. In addition, officials told us that consumer complaints have indicated that service reliability is now as important to consumers as speed due to the popularity of video-streaming services, such as Netflix, and that as a result, they are exploring ways to further incorporate service reliability information into the MBA program. They also told us that based on consumer complaints, they knew that consumers could not easily access—and in some cases were not provided—certain performance information (e.g., ISPs' network management practices, data limits, etc.), and FCC tried to address this issue through the transparency rule.

FCC's Ability to Evaluate Its Efforts Is Limited by a Lack of Useful Performance Information and Relevant Performance Goals and Measures

FCC Obtains Inconsistent and Incomplete Performance Information from Stakeholder Comments and Consumer Complaints

We have previously found that leading organizations that have progressed toward results-oriented management use performance information as a basis for decision-making[33] and that the usefulness of performance information can be affected by its completeness, accuracy, consistency, validity, and credibility, among other things.[34] For example, performance information can focus on various dimensions of performance, such as outcomes, outputs, quality, or customer satisfaction, and can help agencies monitor their progress in meeting programmatic and operational goals. Although stakeholder comments and consumer complaints can provide FCC with valuable insights on topics of interest to the commission, the information FCC obtains from these sources lacks some of the characteristics of useful performance information, such as consistency and completeness.

Our review of selected stakeholders' *Open Internet* comments found that they generally provided inconsistent views regarding the effectiveness of FCC's efforts to provide consumers with broadband performance information. For instance, some public interest groups and content providers argued that FCC's transparency rule was not providing consumers the information they need to make informed decisions regarding broadband services. In their view, FCC should have required ISPs to provide additional technical information—such as detailed information on network congestion—under the transparency rule. However, ISPs countered that this level of technical detail would be confusing to consumers and that the transparency rule had been effective in providing consumers with the information they need. For example, Cox Communications, an ISP, stated that, "most of the expanded disclosure proposals included in the NPRM would impose burdensome obligations that are highly unlikely to prove useful to consumers and would only degrade the overall effectiveness of the transparency rule." FCC has acknowledged that it has not received consistent information in response to its requests for comments. For instance, in its 2010 *Open Internet* Order, FCC states that despite broad agreement that broadband providers should disclose information sufficient to enable end users and content providers to understand the capabilities of broadband services, commenters disagree about the appropriate level of detail required to achieve this goal. As a result, FCC decided to continue to allow ISPs flexibility in the implementation of the transparency rule while providing guidance regarding acceptable disclosure methods. Similarly, in its 2015 *Open Internet* Order, FCC noted that despite its efforts to seek comment on this issue, the record was lacking with respect to specific details on how such a disclosure should be formatted. Thus, as previously mentioned, FCC asked its Consumer Advisory Committee to propose a disclosure format.

As noted above, in the 2014 *Open Internet* NPRM, FCC cited consumer complaints as a basis for tentatively concluding that the transparency rule should be strengthened. However, in the NPRM FCC also acknowledged that its analysis of consumer complaints provided limited insight into the issues that consumers were experiencing. Specifically, FCC stated that since the transparency rule took effect, it had received a "significant number" of consumer complaints about provider speeds, charges, and other practices that the rule was designed to disclose. However, FCC also stated that in some cases, it is difficult to discern from the complaints "whether the consumer's frustration is with slow speeds or high prices generally, or instead with how the service as actually provided differs from what the provider has advertised." Moreover, because consumer complaints reflect the perspective of a non-representative sample of consumers—i.e., those who have experienced broadband issues and have chosen to report those issues to the FCC—they do not fully capture the effectiveness of FCC's efforts for a broader range of consumers. Consequently, several *Open Internet* comments we reviewed filed by ISPs and industry associations questioned FCC's reliance on complaints as a basis for decision-making.

In addition, we have previously noted concerns with FCC's complaint process.[35] In 2009, we found that many wireless phone service consumers were not aware that they could file complaints with FCC when they encountered problems. We also found that FCC did not complete in-depth analyses of consumer complaints and lacked goals and measures that clearly identified the intended outcomes of its complaint-processing efforts. As a result, we noted that FCC may not be aware (1) of emerging trends in consumer problems, (2) if specific rules are being violated, or (3) if additional rules are needed to protect consumers. We made a number of recommendations to help address these issues, and FCC has indicated that it is in the process of reforming the agency's complaint process. For example, in January 2015, FCC launched a new online consumer-help center. According to FCC, the online center will make it easier for consumers to file complaints with FCC, improve communications between consumers and FCC representatives, and help streamline FCC's process for synthesizing and analyzing trends in consumer complaints. While these steps may enhance FCC's ability to assess the effectiveness of its efforts among the subset of consumers who file complaints, they will not provide FCC information regarding the effectiveness of its efforts for a broader range of consumers.

Without consistent and complete information regarding the effectiveness of its efforts to provide consumers with broadband performance information, FCC's ability to evaluate its efforts and make performance-based decisions about these efforts is limited. In addition, FCC may find it difficult to convince stakeholders that enhancements to its transparency rule and MBA program are warranted. As we reviewed industry groups' *Open Internet* comments, we found many stating that FCC did not provide adequate evidence that ISPs' disclosures under the transparency rule were ineffective and that enhancements or changes to the rule were needed. For example, Cisco, an equipment provider, stated that FCC's concerns regarding the accuracy of information some consumers received under the transparency rule do not imply that additional disclosures are called for, but rather that the commission may need to enforce the requirements instead.

In contrast to the approaches FCC typically relies on to evaluate the effectiveness of its efforts to provide consumers with broadband performance information, other government entities have undertaken extensive, qualitative, and quantitative consumer research to evaluate the effectiveness of their efforts to inform consumers and ensure that consumers' information needs are being met. For example:

- Ofcom—the United Kingdom's (UK) communications regulatory agency—has conducted qualitative and quantitative research to determine consumers' use and understanding of ISPs' traffic management information and the effectiveness of a standardized form that all major UK ISPs voluntarily use to publish their traffic management information. For instance, in 2013, Ofcom commissioned a study that involved qualitative workshops with 135 participants and a representative survey of UK broadband consumers. The study found that consumers had problems understanding some aspects of the standardized form and suggested several potential improvements that could make the information more understandable to consumers, such as adding a summary to the form explaining the relevance of traffic management policies to broadband products.
- The U.S. Consumer Financial Protection Bureau (CFPB) conducted qualitative and quantitative consumer research when revamping mortgage disclosure forms. The CFPB tested the prototype of the revised disclosure form through 10 rounds of qualitative testing in nine cities. After issuing the revised forms the CFPB also conducted a large-scale quantitative validation test—that included 858 consumers in 20 locations—to compare the revised forms with the disclosures that were currently in use.

Some stakeholders we spoke with and whose comments we reviewed also suggested that FCC should undertake consumer research to inform and evaluate its efforts to provide consumers with broadband performance information. In 2009, the Federal Trade Commission filed comments with FCC stating that research has shown that well-designed standardized disclosures can improve consumer understanding and facilitate competition; however, to be most effective, these disclosures should be developed and revised based on controlled, quantitative, objective tests of consumer understanding. In addition, representatives from one industry association we spoke with stated that FCC's analytical framework, which largely consists of requesting and reviewing stakeholder comments, is not consumer-centric and may not be the most productive process for obtaining an accurate representation of consumers' information needs. These representatives noted that given the time, effort, and money ISPs expend to collect broadband performance data for FCC's MBA program, it is important to make sure that the data are presented in a way that is meaningful to consumers. In their view, FCC should also be conducting consumer research, such as focus groups, to make sure that its MBA effort produces a report that consumers can understand. According to FCC officials, FCC's goal is to improve the accessibility, accuracy, and relevance of the broadband performance information available to consumers. Because consumers' decision-making processes are complex and can vary, FCC has not attempted to formally evaluate whether its efforts to provide consumers with broadband performance information are meeting consumers' needs. Officials also told us that they do not believe this type of research has been conducted outside of FCC.

FCC Lacks Performance Goals and Measures to Monitor Its Efforts to Inform Consumers

FCC's current strategic plan includes strategic objectives related to informing consumers about broadband networks—reflecting its position that informed consumers help facilitate competition in the broadband market. Specifically, FCC includes "ensure effective policies are in place to promote and protect competition" and "take action where competition is not sufficient to protect the public interest, including ensuring that consumers remain informed" as two of its 17 strategic objectives and states that it will continue to engage consumers though its outreach and education initiatives to facilitate informed choice in the competitive communications marketplace. However, FCC does not include any performance goals under these strategic objectives that define desired outcomes for its efforts to provide consumers with broadband performance information, such as its transparency rule and MBA program. Consequently, FCC also lacks relevant performance measures under these strategic objectives that would allow it to monitor the impact and effectiveness of its efforts. We have previously reported that effective performance goals and measures possess certain characteristics. These include the following: 1) performance goals should clearly define the organization's desired outcomes, and 2) each performance goal should have a few performance measures that clearly tie back to the goal, cover key performance dimensions, and take different priorities into account.36 Without performance goals and measures that allow FCC to monitor its progress on its goals related to providing consumers with broadband performance information, FCC has limited assurance that its efforts to provide consumers with broadband performance information are effective and meeting consumers' needs. In addition, FCC cannot demonstrate that it is achieving its objectives to ensure consumers are informed as envisioned in its strategic plan.

CONCLUSION

In recent years, the broadband market has undergone significant changes. More Americans are accessing broadband and relying on their broadband service for an increasingly diverse range of uses, including education, medicine, public safety, and entertainment. As broadband traffic grows, consumers' broadband performance issues—such as slower than expected speeds and disruptions to video-streaming services—may simultaneously become more prevalent and more critical to resolve. However, broadband performance can be affected by a number of complicated, technical factors that are not readily apparent to most consumers.

FCC has maintained that informed consumers play an important role in facilitating competition in the broadband market and has undertaken efforts to provide consumers with broadband performance information, such as the transparency rule and the MBA program. However, FCC lacks useful performance information and relevant performance goals and measures to evaluate the effectiveness of these efforts and monitor their impact. As a result, FCC cannot be assured that its efforts to provide consumers with broadband performance information are effective and meeting consumers' needs, and its ability to make performance-based decisions will remain limited. Moreover, FCC may find it difficult to convince stakeholders that its enhancements to the transparency rule are warranted and address their concerns that FCC has not provided adequate evidence to support these enhancements.

RECOMMENDATIONS FOR EXECUTIVE ACTION

To help FCC determine whether its efforts to provide consumers with broadband performance information are effective and meeting consumers' needs, and whether additional efforts—such as a standardized label suggested by FCC's transparency working group— could benefit consumers, FCC should take the following two actions:

1. conduct or commission research on the effectiveness of FCC's efforts to provide consumers with broadband performance information and make the results of this research publicly available, and
2. establish performance goals and measures under the agency's relevant strategic objectives that allow it to monitor and report on the impact and effectiveness of its efforts.

AGENCY COMMENTS

We provided a draft of this report to FCC for review and comment. In its written comments, FCC generally concurred with our recommendations. FCC described its various consumer outreach and education efforts and agreed that extensive research would be helpful in evaluating the effectiveness of its efforts to inform consumers. FCC stated that it will evaluate consumer research on broadband performance and move expeditiously to take action. FCC also stated that it would use consumer research, consumer comments, and in-person feedback at consumer-related events to establish performance measures that will allow it to monitor and report on its progress. FCC also provided an appendix with details about its outreach efforts as well as technical comments that were incorporated as appropriate.

Mark Goldstein
Director, Physical Infrastructure

APPENDIX I: OBJECTIVES, SCOPE, AND METHODOLOGY

The objectives of this report were to examine (1) what information is currently available to consumers regarding broadband performance and the limitations of this information, if any, and (2) how does the Federal Communications Commission (FCC) evaluate its efforts to provide consumers with broadband performance information?

To assess what information is currently available to consumers regarding broadband performance and any limitations of this information, we reviewed prior GAO reports on this issue, relevant FCC rulemaking proceedings, including its 2014 *Open Internet* proceeding, FCC's transparency rule and Measuring Broadband America reports, Internet service providers' (ISP) websites, information provided by companies, including Netflix and Google, and third-party sites that provide broadband performance information to consumers. We also conducted a literature review to identify articles and studies relevant to broadband performance measurement and disclosure of broadband performance information to consumers. We used key-word searches to identify peer-reviewed articles and other work in relevant databases and reviewed the websites of researchers identified by FCC as conducting work relevant to our review. We also interviewed representatives from FCC, public-interest and consumer-advocacy groups, industry associations, content or equipment providers, and broadband providers. We identified stakeholders to interview based on our review of comments filed in FCC's 2014 *Open Internet* and related proceedings, as well as based on recommendations from other organizations we interviewed. Finally, we reviewed a judgmental sample of comments filed in FCC's 2014 *Open Internet* proceeding. In this proceeding, FCC requested comment on the effectiveness of its efforts to provide broadband performance information to consumers, including its transparency rule and its Measuring Broadband America program, and its proposed enhancements to these efforts. We ran key-word searches that included terms related to transparency and disclosure for consumers to identify relevant comments, and selected comments from a variety of stakeholders, including public interest groups, ISPs, content providers, trade associations, and consumer representatives. We reviewed these comments to identify statements regarding the effectiveness of FCC's efforts to provide broadband performance information to consumers, including comments supporting and opposing FCC's proposed enhancements to these efforts.

To determine how FCC evaluates its efforts to provide consumers with broadband performance information, we interviewed FCC officials regarding their efforts to provide consumers with broadband performance information and evaluate and monitor the effectiveness of these efforts and reviewed relevant FCC documents, including its strategic and performance plans, consumer surveys, and resources available to consumers on FCC's website.

We also reviewed FCC's requests for comment on consumers' information needs and the effectiveness of FCC's efforts to provide consumers with broadband performance information, as well as selected comments filed in response to these proceedings (as described above). To determine what additional actions, if any, FCC could take to evaluate its efforts to provide consumers with broadband performance information, we interviewed representatives from the stakeholder groups identified above, as well as officials from the Federal Trade Commission and National Telecommunications and Information Administration regarding their efforts to inform consumers about broadband issues and to evaluate the effectiveness of those efforts. In addition, we reviewed reports and studies on consumer disclosures conducted by FCC's Transparency Working Group; foreign telecommunications-regulatory bodies, including those in the United Kingdom and European Union; and other federal agencies, such as the Consumer Financial Protection Bureau. We also reviewed prior GAO reports on FCC and results-oriented management and compared FCC's use of performance information and measures to performance management best practices outlined in previous GAO work.

We conducted this performance audit from July 2014 to April 2015 in accordance with generally accepted government auditing standards. Those standards require that we plan and perform the audit to obtain sufficient, appropriate evidence to provide a reasonable basis for our findings and conclusions based on our audit objectives. We believe that the evidence obtained provides a reasonable basis for our findings and conclusions based on our audit objectives.

End Notes

[1] Includes connections of a download speed of at least 3 megabits per second and an upload speed of at least 768 kilobits per second. Federal Communications Commission, Internet Access Services: Status as of December 31, 2013 (Washington, D.C.: October 2014).

[2] Cisco, Cisco Visual Networking Index: Forecast and Methodology, 2013—2018 (June 10, 2014).

[3] GAO, Telecommunications: Current Broadband Measures Have Limitations, and New Measures Are Promising but Need Improvement, GAO-10-49 (Washington, D.C.: Oct. 9, 2009), and GAO, Broadband Internet: FCC Should Track the Application of Fixed Internet Usage-Based Pricing and Help Improve Consumer Education, GAO-15-108 (Washington, D.C.: Nov. 24, 2014).

[4] Preserving the Open Internet, Report and Order, 25 FCC Rcd. 17905 (2010).

[5] Protecting and Promoting the Open Internet, Notice of Proposed Rulemaking, 29 FCC Rcd. 5561 (2014). Protecting and Promoting the Open Internet, Report and Order on Remand, Declaratory Ruling, and Order, FCC 14-24 (2015).

[6] 47 U.S.C. § 1302(b). Advanced telecommunications capability is defined as "high-speed, switched, broadband telecommunications capability that enables users to originate and receive high-quality voice, data, graphics, and video telecommunications using any technology." 47 U.S.C. § 1302(d).

[7] In January 2015, FCC adopted a speed benchmark at download speeds of at least 25 Mbps and upload speeds of at least 3 Mbps. Inquiry Concerning the Deployment of Advanced Telecommunications Capability to All Americans in a Reasonable and Timely Fashion, and Possible Steps to Accelerate Such Deployment Pursuant to Section 706 of the Telecommunications Act of 1996, as Amended by the Broadband Data Improvement Act, GN Docket No. 14-126, 2015 Broadband Progress Report and Notice of Inquiry on Immediate Action to Accelerate Deployment, FCC 15-10 (rel. Feb. 4, 2015) (2015 Broadband Progress Report) at paras. 3 and 45.

[8] The radio frequency spectrum is the resource that makes possible wireless communications and supports a vast array of commercial and government services.

[9] The ACSI measures the satisfaction that U.S. consumers have with the quality of a number of different types of services, including Internet services. American Customer Satisfaction Index, ACSI Telecommunications and Information Report 2014, (Ann Arbor, MI: May 2014).

[10] In the context of its Measuring Broadband America program (discussed later in this report), FCC defines the peak usage period for residential broadband as weeknights between 7:00 pm to 11:00 pm local time. This is the time when delivery of Internet service is under highest demand.

[11] Mobile service is provided through cell tower coverage with data sharing through radio spectrum; thus, mobile connection quality will vary based on the location of the phone or device in reference to the tower. For example, if the device is blocked by terrain, the mobile broadband service will be degraded or not available. In addition, the performance of the broadband connection degrades over distance to the tower, so that performance at the edge of a tower's coverage is not equal to performance close to the tower.

[12] Cisco estimates that Internet video traffic will grow in North America by 27 percent, from 2013–2018.

[13] A content delivery network deploys many servers in many distributed locations to offer the best quality of experience to the end user. Content delivery networks may build large content distribution centers and connect these centers with high speed links to the ISPs, or they may deploy their cache servers inside the ISPs. Cheng Huang, Angela Wang, Jin Li, and Keith W. Ross, "Understanding Hybrid CDN-P2P: Why Limelight Needs Its Own Red Swoosh," NOSSDAV '08: Proceedings of the 18th International Workshop on Network and Operating Systems Support for Digital Audio and Video (2008): 75-80, cited by Manuel Palacin et.al, "The Impact of Content Delivery Networks on the Internet Ecosystem," Journal of Information Policy, vol. 3 (2013).

[14] Pub. L. No. 111-5. title VI, § 6001, 123 Stat.115 (2009); FCC, Connecting America: The National Broadband Plan (2010)

[15] 47 U.S.C. § 1302(a). As previously noted, advanced telecommunications capability is defined as "high-speed, switched, broadband telecommunications capability that enables users to originate and receive high-quality voice, data, graphics, and video telecommunications using any technology." 47 U.S.C. § 1302(d).

[16] In the Matter of Consumer Information and Disclosure, Notice of Inquiry, 24 FCC Rcd. 11380, ¶5 (2009); Preserving the Open Internet, Report and Order, 25 FCC Rcd. 17905 , ¶53 (2010); Connecting America: The National Broadband Plan, p.44 (2010); In the Matter of Broadband Industry Practices, Notice of Inquiry, 22 FCC Rcd. 7894, ¶2 (2007).

[17] Preserving the Open Internet, Report and Order, 25 FCC Rcd. 17905 (2010), aff'd in part, vacated and remanded in part sub nom, Verizon v. FCC, 740 F.3d. 623 (D.C. Cir. 2014).

[18] Protecting and Promoting the Open Internet, Notice of Proposed Rulemaking, 29 FCC Rcd. 5561 (2014).

[19] Protecting and Promoting the Open Internet, Report and Order on Remand, Declaratory Ruling, and Order, FCC 14-24 (2015).

[20] For example, AT&T's performance disclosure website includes a video explaining the factors that can affect speed, and Cox Communications provides a tool that asks consumers questions about the number of users in their household and what services they use with each device.

[21] 47 C.F.R. § 8.3.

[22] FCC granted a temporary exemption from these enhancements to smaller broadband providers.

[23] FCC's Consumer Advisory Committee previously reviewed pre-sale disclosures for consumers interested in purchasing broadband services. The committee noted that while a label would help consumers compare broadband options prior to purchasing service, a label would be limited in empowering consumers with respect to educating them about which broadband services and speeds would meet their needs. The committee noted that this would be better addressed through a comprehensive educational program and expanded industry disclosure that included FCC, industry, and consumer groups. FCC Consumer Advisory Committee, Recommendations Regarding Pre-sale Consumer Disclosures, GC Docket No 09-158, CC Docket No. 98-170, WC Docket No. 04-36 (Aug. 4, 2010).

[24] FCC, FCC Speed Test App Tip Sheet, accessed February 6, 2015, http://www.fcc.gov /guides/mobile-speed-test-tip-sheet.

[25] FCC's 2014 Measuring Broadband America report noted that as available broadband speeds continue to increase, older modems incapable of supporting such speeds could be a growing problem for consumers' fixed broadband performance. See, FCC, 2014 Measuring Broadband America Fixed Broadband Report (Washington, D.C.).

[26] Some ISPs offer plans that limit how much data a consumer can use per month before receiving slower speeds or paying an additional fee. For example, once customers reach their high-speed data allowance, they may continue to access unlimited data, but at slower connection speeds of 128 kilobits per second or less, slower than speeds FCC recommends for browsing the web or downloading e-mail. See GAO-15-108 for additional information.

[27] FCC, Measuring Broadband America, accessed February 6, 2015, http://www.fcc.gov/ measuring-broadband-america.

[28] FCC, Measuring Broadband America Mobile Broadband Services, accessed February 6, 2015, http://www.fcc.gov/measuring-broadband-america/mobile.

[29] FCC, 2014 Measuring Broadband America Fixed Broadband Report (Washington, DC).

[30] Some ISPs raised concerns regarding comparisons of actual versus advertised speed for broadband delivered via DSL, noting that the actual speed a specific customer receives depends on the distance between the ISP's central office and the customer's home. One ISP we interviewed noted that is why they advertise a speed range, rather than a specific speed.

[31] For example, in 2014, FCC increased, to 10Mbps download/1 Mbps upload speed, the required minimum speed that fixed broadband providers receiving high-cost support must offer. The Universal Service Fund's High Cost program subsidizes telecommunications carriers that

provide basic telephone, broadband and wireless telephone service in generally rural high-cost areas.

[32] According to FCC, this congestion was caused by business disputes between ISPs and a backbone provider. FCC did collect results from the affected servers and released the data for use by academics and others examining such congestion issues.

[33] GAO, Managing for Results: Enhancing Agency Use of Performance Information for Management Decision Making, GAO-05-927 (Washington, D.C.: Sept. 9, 2005).

[34] GAO, Results-Oriented Management: Strengthening Key Practices at FEMA and Interior Could Promote Greater Use of Performance Information, GAO-09-676 (Washington, D.C.: Aug. 17, 2009).

[35] GAO, Telecommunications: FCC Needs to Improve Oversight of Wireless Phone Service, GAO-10-34 (Washington, D.C.: Nov. 10, 2009).

[36] GAO, Executive Guide: Effectively Implementing the Government Performance and Results Act, GAO/GGD-96-118 (Washington, D.C.: June 1, 1996).

In: U.S. Broadband Issues
Editor: Chester Rios

ISBN: 978-1-63483-944-0

Chapter 3

MUNICIPAL BROADBAND: BACKGROUND AND POLICY DEBATE*

Lennard G. Kruger and Angele A. Gilroy

SUMMARY

Since the late 1990s, broadband Internet service has been deployed in the United States, primarily by private sector providers. While broadband deployment has been rapid and robust overall, there remain communities that are dissatisfied with their broadband service. Some of these communities have turned to public entities as possible broadband providers, with the expectation that municipal broadband networks (also referred to as "community broadband") can deliver superior levels of speed, performance, and/or affordability than what is currently offered by private providers. Public entities that provide broadband service can be local governments or public utilities, for example, and may construct and manage broadband networks either solely or in partnership with private companies. There are a number of municipal broadband models that have been implemented across the nation. Since each community is different and each faces unique challenges, there is no one size that fits all.

Municipal broadband is controversial, because it involves governmental entities entering a commercial telecommunications marketplace that had previously been the exclusive domain of private

* This is an edited, reformatted and augmented version of a Congressional Research Service publication R44080, prepared for Members and Committees of Congress, dated August 6, 2015.

sector providers. Supporters of municipal broadband argue that in view of substandard broadband service, communities and local governments should be able to provide this service to meet their citizens' needs and to support the community's economic development. Municipal broadband opponents argue that public entities are ill-equipped to efficiently develop, operate, and maintain commercial broadband networks, and that municipally owned and supported broadband networks constitute unfair competition to private sector providers, which may ultimately impede private investment in broadband infrastructure.

With under 500 municipalities across the nation embarking on some form of municipal broadband, 20 states have passed laws placing restrictions (or in some cases, bans) on local broadband networks. The issue for Congress is whether municipal broadband should be promoted or discouraged, and more specifically, whether those state restrictions on municipal broadband should be overridden or affirmed.

On March 12, 2015, the Federal Communications Commission (FCC) released a Memorandum Opinion and Order granting the petitions filed by two municipal broadband providers in Wilson, NC, and Chattanooga, TN, to preempt state laws in their respective states that restricted the expansion of community broadband services. The Order and the decision by the FCC to rely on Section 706 of the 1996 Telecommunications Act for its authority remain controversial. Both states have filed petitions for review now consolidated in the U.S. Court of Appeals, 6th Circuit, Cincinnati, challenging the FCC's authority to preempt these restrictions.

Three bills (S. 240, S. 597, and H.R. 1106) have been introduced and one draft measure has been released in the 114th Congress addressing the municipal broadband debate. The role of municipal broadband and the appropriate role of the states and the FCC to address the relationship between the public and private sector is just one facet in the overall debate regarding broadband deployment. Whether municipal broadband should be encouraged or restricted is one of the many policies that Congress continues to consider.

BACKGROUND

Broadband—also referred to as high-speed Internet service—has been deployed in the United States since the late 1990s, primarily by private sector providers. These providers include telephone companies, cable companies, wireless providers, and other entities that provide commercial telecommunications services to residential, business, and institutional customers.

While broadband deployment has been rapid and robust overall, there are parts of the nation where broadband is less deployed (primarily rural areas) and there remain regions and communities that are dissatisfied with the level of broadband service currently offered by private sector providers. These communities point to problems ranging from low download and upload speeds, obsolete technology, poor reliability, high prices, and/or a lack of choice in providers. With the Federal Communications Commission (FCC) moving to define the minimum speed of broadband at 25 Mbps,[1] more communities may perceive a lack of adequate broadband service, especially those communities in rural areas.

As a solution, some communities have turned to public entities as possible broadband providers. These communities anticipate that public entities may be able to provide municipal broadband at superior levels of speed, performance, and affordability than what is currently offered by private providers. Public entities that provide broadband service can be local governments or public utilities, for example, and may construct and manage broadband networks either solely or in partnership with private companies. There are a number of municipal broadband models that have been implemented across the nation. Since each community is different and each faces unique challenges, there is no one size that fits all.

Municipal broadband (also sometimes referred to as "community broadband") is a somewhat amorphous term that can signify many different ways that a local government might participate– either directly or indirectly— in the provision of broadband service to the local community. Municipal broadband models can include public ownership, public-private partnership, and a cooperative model.

With *public ownership*, the local government is the principal entity building, financing, and operating the broadband network. The network can be run by the local municipal electric utility (Chattanooga, TN, and Lafayette, LA, are examples), or it can be run by a city department such as the information technology (IT) department (as in Santa Monica, CA). There are also instances where a publicly owned and built network might be opened to private providers to provide retail Internet access or other services to the public.[2]

Public-private partnerships can come in many different forms, from public and private sector entities sharing capital and operations costs, to governments providing access to public rights-ofway or other city infrastructure (e.g., conduits, pole attachments) for privately funded and operated networks, to government funded projects contracting with private providers to build, operate, and/or maintain the network. Partners can include private for-profit companies, local non-profits, and even local residents.[3]

Finally, there is a *cooperative model*, which refers to electric and telephone cooperatives, many of which were originally created during rural electrification in the 1930s. These cooperatives, in rural areas, have begun in some instances to provide broadband service. Many of the cooperatives providing broadband service have received or are eligible for federal loan and grant support from the Rural Utilities Service (RUS) of the U.S. Department of Agriculture (USDA). There are also a few cooperatives that have been recently formed specifically for providing broadband service. These typically rely on support from local governments and include the East Central Vermont Community Fiber Network (ECFiber) and the WiredWest project in western Massachusetts.[4]

Aside from the models of how municipal broadband networks are governed, the nature of broadband service offered by municipal broadband networks can vary. Municipal networks

- may provide wholesale service ("middle-mile" infrastructure, where retail providers connect into the municipal network), "last mile" retail service directly to customers, or both;
- may provide service solely to anchor institutions or also include businesses and residences;
- may serve solely within municipal boundaries or may be extended to surrounding municipalities and counties;
- may provide data or data bundled with video and/or voice, or may include smart grid capacity; and
- while most recent and proposed municipal broadband projects utilize fiber infrastructure, other broadband technologies such as wireless or cable have also been deployed.

MUNICIPAL BROADBAND NETWORKS IN THE UNITED STATES

Municipal broadband networks tend to be established in small and mid-sized communities, often located in rural areas. With some exceptions, municipal broadband networks are typically not located in major metropolitan areas, where many private providers already offer broadband service.

The Institute for Local Self-Reliance (ILSR) lists 492 U.S. municipalities with broadband networks. The complete list is included in the appendix of the January 2015 White House report *Community-Based Broadband Solutions: The Benefits of Competition and Choice for Community Development and Highspeed Internet Access.*[5] This includes 89 communities with a publicly owned fiber to the home (FTTH) network reaching most or all of the community, 76 communities with a publicly owned cable network reaching most or all of the community, over 180 communities with some publicly owned fiber service available to parts of the community, over 110 communities with publicly owned dark fiber available, and over 40 communities in 13 states with a publicly owned network offering at least 1 gigabit services.[6]

The magazine *Broadband Communities* lists 143 public and public-private fiber to the premises (FTTP) network projects.[7] This list identifies community fiber systems in 37 states and in American Samoa. The largest number of deployments is in Washington (13), Kentucky (11), Minnesota (10), Tennessee (8), Iowa (8), Illinois (7), and Florida (7).[8]

POLICY DEBATE: PROS AND CONS

The virtues and drawbacks of municipal broadband have been vigorously debated by policymakers and other stakeholders.[9] Advocates for municipal broadband include groups aligned with local communities, while opponents include private sector incumbent broadband providers and state governmental entities.

Arguments in Favor of Municipal Broadband

The primary argument in favor of municipal broadband is rooted in the dissatisfaction of some communities with existing broadband service that is offered by private providers. Many local communities cite low speeds, high prices, a lack of competition, or even an absence of any broadband service in particularly sparsely populated areas, and argue that they should be able to provide this service to meet their citizens' needs and to support the community's economic development. Pro-municipal broadband arguments include

- Municipal broadband can enable small and mid-sized municipalities, often in rural areas, to offer higher download and upload speeds. This is especially important given that the FCC continues to identify a persistent "digital divide" between rural areas (where 53% of Americans do not have broadband speeds of at least 25 Mbps download/3 Mbps upload[10]) and urban areas (where only 8% do not have access to those speeds).[11] Overall, 16% of American households are in areas without a single provider of 25 Mbps/3 Mbps fixed services.[12]
- Municipal networks can inject competition in markets where there may be a limited number of providers. According to the FCC, 45% of households have only a single provider of broadband offering 25 Mbps/3 Mbps. A lack of competition can lead to high prices, poor customer service, limited and restrictive service packages, and delayed or no investment in advanced technologies such as ultra-fast gigabit networks. A municipal broadband network, in some cases, can induce private providers to lower prices and increase speeds in order to compete.
- Municipal broadband can address unmet public interest needs. Private providers tend to favor middle- to upper-income households which will generate adequate revenue. Municipal broadband entities that are publicly owned may be more likely to offer broadband to low-income households at affordable prices.
- Municipal broadband follows the tradition of municipal utilities, which have been providing basic utilities such as water, natural gas, and electricity for many years.

Arguments Opposed to Municipal Broadband

The main argument against municipal broadband (typically referred to by some opponents as "government-owned broadband") is that it is inappropriate for government-sponsored, -owned, or -supported networks to compete with private providers. Municipal networks have unfair inherent advantages over existing private networks, including preferential treatment with respect to rights of way and other local regulatory barriers, and financing by direct taxpayer subsidies or government bonds with below-market interest rates. This advantage can result in market-distorting effects that can unfairly skew the competitive playing field between private and public providers. Anti-municipal broadband arguments include

- Deploying broadband systems are inherently high-risk, because unlike basic utilities like water or electricity, there are typically competing providers and not all customers will necessarily sign up for service. Governments can be ill-equipped to plan, operate, and maintain efficient commercial broadband systems, and if they fail, the taxpayers will be liable for the cost of that failure.
- Taxpayer money should more appropriately be directed towards basic infrastructure needs—such as roads, bridges, and water systems—that are traditionally under the purview of government. In the United States, broadband is primarily provided by the private sector. Public money that is directed towards municipal broadband is money that is taken away from other, more critical infrastructure needs.
- According to the FCC, "private industry continues to invest billions of dollars to expand America's broadband networks."[13] Because of the market-distorting effects of municipal broadband, continued private sector investment in broadband networks might be discouraged in some cases.
- The broadband market is subject to rapid technological change and intense competition. The bureaucracy of government is not well suited to making policy decisions in a dynamic and rapidly changing environment. This poses the risk of municipal broadband networks being reliant on soon-to-be obsolete technologies.

Case Studies—Successes or Failures?

While all agree that there is risk in municipal broadband, supporters and opponents argue over the significance of "successes" and "failures" among existing municipal broadband projects. With hundreds of municipal broadband projects to choose from, there will always be examples to fit whichever definition of "success" or "failure" that observers choose to apply. In general, municipal broadband supporters point to projects that have provided improved services, lower prices, increased competition, and an improved climate for private-sector investment in the local economy.[14] Municipal broadband opponents cite examples where government-owned networks have not been profitable, have discouraged private competition, and have been subject to managerial inefficiency or technological obsolescence.[15]

In some cases, both proponents and opponents of municipal broadband have cited the same municipal broadband project to bolster their arguments. For example, in 2005, the community of Lafayette, LA, voted to build a municipal fiber network called LUS Fiber. LUS Fiber, financed by bond revenues, was built in 2008 and connected to its first customers in 2009.

According to the White House report *Community-Based Broadband Solutions: The Benefits of Competition and Choice for Community Development and Highspeed Internet Access*, LUS Fiber's network has increased customer savings and strengthened local anchor institutions:

> As competing firms adjusted their plans to account for LUS Fiber's market entry, residents who weren't customers of the network started to see lower prices. Cox Communications, a major regional provider which had raised rates six times in four years, kept its rates stable from 2004 to 2007 to account for LUS's possible market entry. Still, LUS's prices have been consistently lower than those offered by Cox. Terry Huval, the director of LUS, estimates that the community saved $4 million from these deferred rate increases. Using estimates of Cox's average competing discounts and LUS Fiber's lower rates, LUS projects the fiber system will create total savings of between $90 and $100 million over its first 10 years.
>
> The fiber network has brought in companies eager to obtain fast service at lower prices. Pixel Magic brought 100 to 200 jobs when it built an office in Lafayette to accomplish work on the movie "Secretariat." The high speed capability of the broadband network was a big factor in their eventual decision to maintain their office in Louisiana permanently. The tech startup firm Skyscraper Holding moved from Los Angeles to Lafayette to obtain 100 Mb/s speeds at a fraction of the cost the company was charged on the west coast.[16]

Municipal broadband opponents have a different take on LUS Fiber, stating the network is 30% short of its revenue projection as set out in its business plan, more than $160 million in debt, and struggling to compete with cable, telephone, wireless, and satellite service providers in terms of price, performance, and service options.[17] The think tank R Street noted that LUS Fiber received a warning from city auditors about low revenues and stated:

> Lafayette's auditors voiced similar concerns in their reports the last two years. In 2012, they punctuated it with a calculation that the $140-million system was costing the city $45,000 a day.

Now, after six years of operation, prospects aren't much better. The city's financial reports, provided by a source in Lafayette, show that for the fiscal year ended Oct. 31, 2013, LUS Fiber reported $23 million in operating revenues, compared to $36.7 million that was forecast in its feasibility study. The system incurred a $2.5 million operating loss for the year. According to the original plan, this was to be the point where the operation swung to a profit of $902,000. The most staggering number, however, is LUS Fiber's deficit, which stood at $47 million at the end of October, up from $37.1 million the year before.[18]

THE ROLE OF THE FEDERAL COMMUNICATIONS COMMISSION

The FCC, an independent federal agency charged with regulating interstate and international communications, has taken an active role in promoting the deployment of broadband services and broadband infrastructure.[19] The FCC has adopted numerous proceedings to facilitate access to and the adoption of advanced services including the following:

- the transition of the Universal Service Fund from a mechanism that supports voice service to one that supports the deployment and adoption of both fixed and mobile broadband;
- the modernization of the Schools and Libraries Program to incorporate high speed broadband and Wi-Fi connections; and
- the proposed expansion of the Low Income Program to provide support for broadband as well as voice services, to name a few.

FCC Chairman Wheeler has also stated on numerous occasions his support for the development of community-based broadband service options and has expressed his opinion that the FCC has the authority to preempt state laws that ban competition from community broadband.[20]

The City of Wilson and the Power Board of Chattanooga Petitions

On July 24, 2014, two local municipally owned broadband providers, the City of Wilson (Wilson), an NC municipal corporation, and the Electric Power

Board of Chattanooga (EPB), an independent board of the City of Chattanooga, TN, separately petitioned the FCC to preempt certain provisions of their respective states' laws which they claimed restricted the further deployment of their networks.[21]

Both Wilson and EPB operate electric utilities that also offer gigabit speed broadband networks that provide data, video, and voice services. Wilson provides electric service in six counties in eastern North Carolina and broadband service solely in Wilson County.[22] Wilson claims that despite "... numerous requests for these services ... in the other five counties.... " and a willingness to expand broadband services to these counties, it cannot, due to what it stated are overly burdensome provisions in state law that in effect have "... the purpose and effect of prohibiting it from doing so."[23] As in the case of Wilson, EPB states that it regularly receives requests from citizens and businesses, located outside of EPB's electric service territory, to provide advanced telecommunications services (e.g., broadband Internet access and services). EPB states that it is willing to provide these services and expand its service footprint, but is restricted by Tennessee state law that permits authorized municipal electric systems to provide Internet service (as well as cable service and video), but only within the boundaries of their (electric) service areas.[24] Both petitioners allege that existing provisions in their respective states' laws restricted their ability to expand their broadband services to surrounding areas where customers have expressed interest in these services and both request that the FCC use its authority pursuant to Section 706 of the Telecommunications Act of 1996 to preempt these laws.[25]

The FCC Memorandum Opinion and Order

The FCC's Wireline Competition Bureau released a public notice on July 28, 2014, establishing a pleading cycle for the petitions setting comment and reply dates of August 29, 2014, and September 29, 2014, respectively.[26] After consideration of record the FCC, in a February 26, 2015, action, granted the petitions to preempt state laws in North Carolina and Tennessee that restricted the expansion of community broadband services.[27] In a Memorandum Opinion and Order (Order), which became effective upon its release on March 12, 2015, the FCC stated that selected provisions of the laws in North Carolina and Tennessee are barriers to broadband deployment, investment, and competition, and conflict with the FCC's mandate to promote these goals.[28]

The FCC relied upon its authority under Section 706 of the 1996 Telecommunications Act (Section 706), which directs the FCC to "... encourage the deployment on a reasonable and timely basis of advanced telecommunications capability to all Americans ... by utilizing ... measures that promote competition in the local telecommunications market, or other regulating methods that remove barriers to infrastructure investment."[29] According to the Order the FCC concludes that "... preemption meets the standard for action under Section 706 because it will remove barriers to overall broadband infrastructure investment and promote overall competition in the telecommunications market in Tennessee and North Carolina."[30] Furthermore, the Order stated that "... preemption of these restrictions will expand broadband investment and deployment, increase competition, and serve the public interest, as Section 706 intended."[31]

The FCC preempted the geographic restrictions of both the Tennessee and North Carolina laws stating that they are barriers to broadband infrastructure investment and competition and preempted additional provisions of the North Carolina law containing other limitations, stating that the cumulative effect of those provisions collectively amount to a barrier to broadband investment and competition.[32] These barriers are in clear conflict, the Order states, with Section 706, which directs the FCC to take action to remove such barriers. More specifically the Order concludes that in the case of the EPB petition "the territorial restriction in Tennessee Code Section 601 is a barrier to broadband deployment and infrastructure investment and limits competition."[33] With regard to the Wilson petition, the Order concludes that the geographic restrictions[34] and other, but not all of the remaining provisions of North Carolina law cited in the petition, when considered holistically, represent a barrier to broadband infrastructure investment or thwart competition.[35] Therefore the Wilson petition is granted in part to the extent discussed in the Order and otherwise denied.[36]

The FCC stated that while it believes it cannot preempt state laws that outright ban municipal broadband networks, it can intervene (under the authority contained in Section 706) if a state allows municipal broadband networks, but imposes restrictions that create barriers to a timely and reasonable deployment of advanced telecommunications services to all Americans. That is, the FCC cannot require a state to allow municipal broadband networks, but it can preempt laws that impose restrictions on an existing network if they are creating barriers to deployment of such networks.[37] While the Order states that this ruling only applies to provisions of the laws of the two states (North Carolina and Tennessee) of the two petitioners, the FCC noted that "... the Commission [FCC] will not hesitate to preempt similar statutory provisions in factual situations where they function as barriers to broadband investment and competition."[38]

Whether the FCC does, or does not, have the legal authority under Section 706 to preempt state laws that restrict municipal broadband deployment remains controversial.[39] While the majority of the FCC commissioners (Chairman Wheeler and Commissioners Clyburn and Rosenworcel) voted in favor of this decision, it was not unanimous. Both Commissioner Pai and Commissioner O'Rielly dissented, stating that the FCC lacked the authority to grant the petitions.[40]

Both the state of Tennessee and the state of North Carolina have filed lawsuits (petitions for review) challenging the FCC's authority to preempt these restrictions. The state of Tennessee filed its petition on March 20, 2015, with the U.S. Court of Appeals 6th Circuit, Cincinnati.[41] The state of North Carolina filed its petition on May 11, 2015, with the U.S. Court of Appeals 4th Circuit, Richmond.[42] These petitions were consolidated on August 3, 2015, in the U.S. Court of Appeals 6th Circuit, Cincinnati and a briefing schedule set.[43]

ADMINISTRATION AND FCC INITIATIVES

In January 2015, President Obama announced steps "to help more Americans, in more communities around the country, get access to fast and affordable broadband."[44] In addition to supporting the FCC Order (discussed above), the Administration plan contained several initiatives directly relevant to municipal broadband, including the following.

- *Establishment of the Broadband Opportunity Council.* On March 23, 2015, the President signed a Presidential Memorandum, "Expanding Broadband Deployment and Adoption by Addressing Regulatory Barriers and Encouraging Investment and Training."[45] The memorandum established an interagency Broadband Opportunity Council which will be chaired by the Department of Commerce (DOC) and the USDA, and consist of 25 other member agencies. The Council's objectives are to engage with industry and other stakeholders to understand ways the government can better support the needs of communities seeking to expand broadband access and adoption; identify regulatory barriers unduly impeding broadband deployment, adoption, or competition; survey and report back on existing programs that currently support or could be modified to support broadband competition, deployment, or adoption; and take all necessary actions to remove these barriers and realign existing programs to increase broadband competition, deployment, and adoption. On April 29, 2015, DOC and USDA put out a notice and request for public comment in the *Federal Register*.[46] The Council is supposed to submit a final report to the President in August 2015.[47]

- *BroadbandUSA*. Based on the expertise acquired from administering the American Recovery and Reinvestment Act of 2009 (ARRA, P.L. 111-5) broadband stimulus program (specifically the Broadband Technology Opportunities Program), the National Telecommunications and Information Administration (NTIA) has established an information and best-practices resource available to communities seeking to develop broadband public-private partnerships.[48] BroadbandUSA[49] will offer online and in-person technical assistance to communities; host a series of regional workshops around the country; and publish guides and tools[50] intended to help communities address problems in broadband infrastructure planning, financing, construction, and operations across many types of business models.
- *Community Broadband Summit*. The White House announced that it will convene a Community Broadband Summit where city and county officials will share information and experiences in deploying broadband for their communities. On March 23, 2015, the White House announced it will host the Community Broadband Summit in June 2015.[51] To date, the scheduling of the Community Broadband Summit has not been announced.

The federal government has also affected municipal broadband through broadband funding programs. While municipal broadband projects are locally directed and funded, the federal government has supported these efforts by helping to finance some of the middle-mile fiber networks that municipal networks can interconnect with. A major funding vehicle for middle-mile fiber networks was the $7 billion broadband stimulus program established by the ARRA.[52] ARRA Awards were made in FY2009 and FY2010, and projects are completed or in the final stages of completion.[53] Going forward, the ARRA broadband programs have concluded and no more funding will be awarded.

Currently, there are three ongoing programs at the RUS that provide funding for broadband infrastructure (although at funding levels significantly less than what was provided in the ARRA broadband programs). These are:

Farm Bill Broadband Loans and Loan Guarantees[54] ($24 million loan level in FY2015), Telecommunications Infrastructure Loans and Loan Guarantees[55] ($690 million loan level yearly), and Community Connect Grants[56] ($10 million in FY2015). While local governmental entities are eligible to apply for these programs, funding has tended to go to private providers.

The other major existing federal vehicle for funding broadband infrastructure is the Connect America Fund (CAF). While RUS grants and loans are used as up-front capital to invest in broadband infrastructure, the CAF provides ongoing subsidies to keep the operation of broadband networks in high-cost areas economically viable for providers.

CONGRESSIONAL ACTIVITY—114TH CONGRESS

Three bills (S. 240, S. 597, and H.R. 1106) have been introduced, and one draft measure (H.R.) released, in the 114th Congress that address the municipal broadband debate. Provisions in these measures range from those that restrict states and localities from enacting laws that prohibit public (municipal) broadband (S. 240) to those that prevent the FCC from preempting current or future state and local laws that prohibit municipal broadband (S. 597 and H.R. 1106), and in the case of the discussion draft (H.R.), preempt the FCC and/or any state regulatory authority from using Section 706 as a source of authority to preempt state laws (e.g., those that prohibit municipally owned broadband networks).

The Community Broadband Act of 2015 (S. 240), introduced by Senator Booker on January 22, 2015, seeks to remove state barriers for constructing municipal broadband networks and encourages public-private partnerships. S. 240 provides that no state or local statute may prohibit, or have the effect of prohibiting or substantially prohibiting, any public entities from providing either telecommunications services (e.g., telephone services) or advanced telecommunications capability or services (e.g., broadband Internet access services). With respect to the private provider that a municipality regulates, S. 240 requires a public provider not to discriminate in favor of its own public network with respect to how it applies municipal ordinances, rules, policies, and fees related to requirements such as rights of way and permitting. S. 240 encourages public-private partnerships and requires extensive public notice of proposed municipal broadband projects, including an opportunity for private providers to bid on that proposed project. The antidiscrimination and public notice requirements in the bill would not apply where a public provider does not provide telecommunications or broadband services to the public "or to such classes of users as to make the capability or services effectively available to the public," or during an emergency. S. 240 prohibits the use of federal funds to assist a public provider in reviving or renewing a project that has failed due to bankruptcy or termination. The bill was referred to the Senate Committee on Commerce, Science, and Transportation.

The State's Rights Municipal Broadband Act of 2015 (S. 597 and H.R. 1106), introduced on February 26, 2015, by Senator Tillis and Representative Blackburn, respectively, states that the FCC cannot preempt states with municipal broadband laws already in place, or any other states that subsequently adopt such municipal broadband laws. The bill also includes a Sense of Congress stating that the FCC does not have the legal authority under Section 706 to prohibit states from implementing any law of such state with respect to the provision of broadband Internet access service (e.g., municipal broadband restrictions). The bills were referred to the House Committee on Commerce, Science, and Transportation and the House Subcommittee on Communications and Technology, respectively.

Draft legislation[57] released on January 16, 2015, by Republican leaders of the House Energy and Commerce Committee and the Senate Committee on Commerce, Science, and Transportation includes a provision that prohibits the FCC, or any state commission with regulatory authority over telecommunications services, from relying on Section 706 as a grant of authority. If enacted this would be in direct conflict with the FCC's final Order which rests on its Section 706 authority to preempt selected provisions of North Carolina and Tennessee law that restrict municipal broadband deployment.

POLICY ISSUES

Since the private sector began deploying broadband infrastructure in the late 1990s, Congress and the FCC have sought to enact policies and programs that address the directive of Section 706 to "encourage the deployment on a reasonable and timely basis of advanced telecommunications capability to all Americans."

With respect to municipal broadband, the issue for Congress is whether locally owned and/or supported networks should be encouraged or restricted. The debate is complicated by the diversity of municipal broadband projects. Each community and project is unique and subject to different factors that can lead to its ultimate success or failure. Abundant examples of successes and failures are available to support arguments made by both supporters and opponents alike.

In addressing municipal broadband, Congress and the FCC have sought to balance two competing public policy interests. On the one hand, with hundreds of municipal broadband projects underway in communities across the country, with other communities exploring various kinds of municipal networks that might offer higher speeds at affordable prices, and with 20 state laws that ban or restrict municipal broadband projects, many have argued that state restrictions be overridden either by congressional legislation or by FCC rule. Ultimately, as discussed above, on March 12, 2015, the FCC released an order lifting restrictions on municipal broadband networks in Wilson, NC, and Chattanooga, TN.

On the other hand, counterbalancing arguments point to the primacy of private sector providers in deploying the nation's broadband. Municipal broadband opponents argue that public entities are ill-equipped to efficiently develop, operate, and maintain commercial broadband networks, and that municipally owned and supported broadband networks constitute unfair competition to private sector providers, and may ultimately impede private investment in broadband infrastructure.

One way that Congress has addressed the debate is through its oversight and authorization of the FCC. Committees with jurisdiction over telecommunications policy—such as the House Energy and Commerce Committee and the Senate Commerce, Science and Transportation Committee— are considering measures reflecting both sides of the issue: from preventing the FCC from overruling state municipal broadband restrictions on the one hand, to overriding those state imposed restrictions on the other.

Congress can also have an impact through the appropriations process. For example, in the 113th Congress, H.R. 5016 (Financial Services and General Government Appropriations Act, 2015), as passed by the House on July 16, 2014, would have provided that none of the funds made available in the FY2015 FCC appropriation could be used to prevent 20 states from implementing their own laws with respect to the provision of broadband by the state or a municipality or other political subdivision of the state.[58]

Another way Congress could support municipal broadband is through funding broadband infrastructure, although funding initiatives are often balanced against fiscal considerations and against concerns over whether federally funded networks unfairly compete against private sector broadband deployment.

Ultimately, whether municipal broadband should be encouraged or restricted is one of many policies that Congress continues to consider for promoting broadband deployment. These include loans and grants for broadband infrastructure deployment; universal service reform; tax incentives to encourage private sector deployment; regulatory and deregulatory measures; and spectrum policy to spur roll-out of wireless broadband services. Some of these policies may be considered in the context of efforts to rewrite the Communications Act of 1934. To the extent that Congress may consider the various options for promoting broadband, a central issue is how to strike a balance between providing government support for broadband in areas where the private sector may not be providing acceptable levels of broadband service, while at the same time minimizing any deleterious effects that government intervention in the marketplace may have on competition and private sector investment.

End Notes

[1] See Federal Communications Commission, 2015 Broadband Progress Report, available at https://www.fcc.gov/ reports/2015-broadband-progress-report.

[2] New America Foundation, The Art of the Possible: An Overview of Public Broadband Options, May 6, 2014, p.8, available at https://static.newamerica.org/attachments/197-the-art-of-the-possible-an-overview-of-public-broadbandoptions/TheArtofthePossible-OverviewofPublicBroadbandOptions_NAFOTI-CTC.pdf.

[3] Ibid., p. 10.

[4] Ibid., pp. 12-13.

[5] Executive Office of the President, Community-Based Broadband Solutions: The Benefits of Competition and Choice for Community Development and Highspeed Internet Access, January 2015, pp. 20-33, available at http://www.whitehouse.gov/sites/default/files/ docs/community- based_broadband_report_by_executive_office_of_the_president.pdf.

[6] Institute for Local Self-Reliance, "Community Network Map," updated January 2015, available at http://www.muninetworks.org/communitymap.

[7] Zager, Masha, "Number of Community FTTP Networks Reaches 143," Broadband Communities, August/September 2014, pp. 10-22, available at http://www.bbpmag.com/ http://www.bbpmag.com/2014mags/Aug_Sep/ BBC_Aug14_CommunityNetworks.pdf.

[8] Ibid., p. 14.

[9] Reports supporting municipal broadband include Executive Office of the President, Community-Based Broadband Solutions: The Benefits of Competition and Choice for Community Development and Highspeed Internet Access, January 2015; and New America Foundation, The Art of the Possible: An Overview of Public Broadband Options, May 6, 2014. Reports opposing municipal broadband include Advanced Communications Law & Policy Institute, New York Law School, Understanding the Debate over Government-Owned Networks, June 2014; and Coalition for the New Economy, The Hidden Problems with Government-Owned Networks, January 6, 2012.

[10] In its 2015 Broadband Progress Report, the FCC raised its broadband speed benchmark from 4 Mbps (download)/1 Mbps (upload) to 25 Mbps/3 Mbps. This benchmark speed upgrade is controversial. The FCC argues that 25 Mbps/3 Mbps is reflective of advanced telecommunications capability, while many providers assert that the new benchmark is too high, excessive, or aspirational. See Federal Communications Commission, 2015 Broadband Progress Report, FCC 15-10, February 4, 2015, pp. 29-34.

[11] Federal Communications Commission, 2015 Broadband Progress Report, FCC 15-10, February 4, 2015, p. 4, available at https://www.fcc.gov/reports/2015-broadband-progress-report.

[12] Ibid., p. 47.

[13] Ibid., p. 9.

[14] See for example: Executive Office of the President, Community-Based Broadband Solutions: The Benefits of Competition and Choice for Community Development and Highspeed Internet Access, January 2015, pp. 13-18; Institute for Local Self-Reliance, Community Broadband Networks, "Successes and Failures," available at http://www.muninetworks.org/content/successes-and-failures; and Edward Wyatt, "Fast Internet is Chattanooga's New Locomotive," New York Times, February 3, 2014.

[15] See for example: Thomas A. Schatz and Royce Van Tassell, "Municipal Broadband Is No Utopia," Wall Street Journal, June 19, 2014; Free State Foundation, "Another One Bites the Dust: Burlington Telecom's Failure Shows, Again, That Government-Operated Broadband Networks Are Not The Solution," March 3, 2014, available at http://freestatefoundation.blogspot.com/2014/03/another-one-bites-dust-burlington.html; and George S. Ford, Phoenix Center for Advanced Legal & Economic Public Policy Studies, "Why Chattanooga Is Not the 'Poster Child' for Municipal Broadband," January 20, 2015, available at http://www.phoenix-center.org/perspectives/Perspective15- 01Final.pdf.

[16] Community-Based Broadband Solutions: The Benefits of Competition and Choice for Community Development and Highspeed Internet Access, p. 16.

[17] Steven Titch, Reason Foundation, Lessons in Municipal Broadband from Lafayette, Louisiana, November 2013, pp. i-ii, available at http://reason.org/files/municipal_ broadband_ lafayette.pdf. A rebuttal was published by Christopher Mitchell, MuniNetworks.org, and Institute for Local Self-Reliance, Correcting Community Fiber Fallacies: The Reality of Lafayette's Gigabit Network, September 2014, available at http://ilsr.org/wp-content/uploads/downloads/2014/10/ fiberfallacieslusfiber.pdf.

[18] Titch, Steven, R Street, Muni broadband: The Gift That Keeps on Taking, May 30, 2014, available at http://www.rstreet.org/2014/05/30/muni-broadband-the-gift-that-keeps-on-taking/.

[19] For a further discussion of the structure and role of the FCC see CRS Report RL32589, The Federal Communications Commission: Current Structure and Its Role in the Changing Telecommunications Landscape, by Patricia Moloney Figliola.

[20] For example, see the statement of FCC Chairman Wheeler before the House Subcommittee on Communications and Technology, May 20, 2014, hearing on "Oversight of the Federal Communications Commission," available at https://apps.fcc.gov/edocs_public/attachmatch/DOC-327165A1.pdf. However, it should be noted that this opinion is not universally held by all of the FCC Commission members.

[21] Petition of the City of Wilson, North Carolina, Pursuant to Section 706 of the Telecommunications Act of 1996, for Removal of Barriers to Broadband Investment and Competition, filed July 24, 2014, available at apps.fcc.gov/ecfs/document/view?id=7521737310.

Petition of the Electric Power Board of Chattanooga, Tennessee, Pursuant to Section 706 of the Telecommunications Act of 1996, for Removal of Barriers to Broadband Investment and Competition, filed July 24, 2014, available at https://www.epb.net/downloads/legal/EPB-FCC petition.pdf.

[22] Wilson County qualified for a grandfathering exemption since it provided these services in that county prior to the law's passage, but is not permitted to provide these services in the five other counties in which it provides electric service.

[23] Wilson petition, p. 2. (N.C. Gen. Stat. sec. 160a-340 through 160a-340.6).

[24] EPB petition, p. 16. (Tenn. Code Ann. sec. 7-52-601).

[25] Section 706 of the Telecommunications Act of 1996, P.L. 104-104, sec. 706, 110 Stat. 56, 153 (1996), as amended by the Broadband Data Improvement Act, P.L. 110-385, 122 Stat. 4096 (2008), is now codified in Title 47, Chapter 12 of the United States Code, at 47 U.S.C. Sec. 1302.

[26] Pleading Cycle Established for Comments on Electric Power Board and City of Wilson Petitions, Pursuant to Section 706 of the Telecommunications Act of 1996, Seeking Preemption of State Laws Restricting the Deployment of Certain Broadband Networks, WC Docket Nos. 14-115 and 14-116, Public Notice, DA 14-1072 (Wireline Comp. Bur. rel. July 28, 2014). Available at http://apps.fcc.gov/eefs/document/view?id=7521737783.

[27] The 3-2 vote fell along party lines with Chairman Wheeler and Commissioners Clyburn and Rosenworcel approving and Commissioners Pai and O'Rielly dissenting. Statements of the Chairman and the commissioners available at https://apps.fcc.gov/edocs_public/attachmatch/FCC-15-25A1.pdf.

[28] In the Matter of City of Wilson, North Carolina, Petition for Preemption of North Carolina General Statute Sections 160A-340 et seq., and the Electric Power Board of Chattanooga, Tennessee, Petition for Preemption of a Portion of Tennessee Code Annotated Section 7-52-601, WC Docket No. 14-115 and WC Docket No. 14-116. Released March 12, 2015. Available at https://apps.fcc.gov/edocs_public/attachmatch/FCC-15-25A1.pdf.

[29] The Telecommunications Act of 1996, P.L. 104-104, Sec. 706.

[30] Order, at para. 42.

[31] Order, at para. 15.

[32] The EPB petition is granted and the Wilson petition is granted in part, Order, paras 183 and 184. The FCC determined that not every provision of the North Carolina law represents a barrier to infrastructure investment or thwarts competition such that they felt compelled to preempt it, para. 182.

[33] Order at para. 168 and paras. 77-79.

[34] As stated in the Order "... restrictions on the provision of bundled services undermines a provider's ability to provide broadband successfully due to a strong customer preference for bundled offerings." Order, para. 119.

[35] Order, at paras.94 and 182.

[36] See Order, paras. 123-129 for the justification for not preempting selected provisions and paras. 123 and 182 for a list of the provisions of North Carolina law not preempted.

[37] Order, at paras. 11, 162-167.

[38] Order, at para. 16.

[39] The legality of this action goes beyond the scope of this report. For a discussion of the legal issues regarding municipal broadband and federal preemption see CRS Legal Sidebar, Municipal Broadband and Federal Preemption, by Kathleen Ann Ruane.

[40] Dissenting Statement of Commissioner Ajit Pai and Dissenting Statement of Commissioner Michael O'Rielly. Available at https://apps.fcc.gov/edocs_public/attachmatch/FCC-15-25A1.pdf.

[41] Tennessee v. FCC, Case No. 15-3291 (6th Cir.) petition for review filed March 20, 2015.

[42] North Carolina v. FCC, Case No. 15-1506 (4th Cir.) petition for review filed May 11, 2015.

[43] Copies of the Order and briefing schedule are available at https://prodnet.www.neca.org/publicationsdocs/wwpdf/ 8315order2.pdf and https://prodnet.www.neca.org/publications docs/wwpdf/8315briefletter.pdf.

[44] The White House, Fact Sheet, "Broadband That Works: Promoting Competition & Local Choice in Next-Generation Connectivity," January 13, 2015, available at https://www.whitehouse.gov/the-press-office/2015/01/13/fact-sheetbroadband-works-promoting-competition-local-choice-next-gener.

[45] Available at https://www.whitehouse.gov/the-press-office/2015/03/23/presidential-memoran dum-expandingbroadband-deployment-and-adoption-addr.

[46] Department of Commerce and Department of Agriculture, "Broadband Opportunity Council Notice and Request for Comment," Federal Register, Vol. 80, No. 82, April 29, 2015, pp. 23785-23787, available at http://www.ntia.doc.gov/ files/ntia/publications/fr_boc _notice_and_rfc_4-29-15.pdf.

[47] Department of Commerce and Department of Agriculture, "Broadband Opportunity Council (BOC) Request for Comments," Webinar, May 20, 2015, p. 5, available at http://www.ntia.doc.gov/files/ntia/publications/ boc_rfc_slides_2015_05_20_final.pdf.

[48] Lawrence E. Strickling, Administrator, National Telecommunications and Information Administration, "NTIA Announced BroadbandUSA Effort to Assist Communities with Broadband Plans," January 14, 2015, available at http://www2.ntia.doc. gov/ntia_announces_broadbandusa_effort.

[49] Available at http://www2.ntia.doc.gov/.

[50] See for example, NTIA, BroadbandUSA: An Introduction to Effective Public-Private Partnerships for Broadband Investments, January 2015, 16 p., available at http://www 2.ntia.doc.gov/files/ntia_ppp_010515.pdf.

[51] The White House, Fact Sheet, "Next Steps in Delivering Fast, Affordable Broadband," March 23, 2015, available at https://www.whitehouse.gov/the-press-office/2015/03/23/fact-sheet-next-steps-delivering-fast-affordable-broadband.

[52] See CRS Report R41775, Background and Issues for Congressional Oversight of ARRA Broadband Awards, by Lennard G. Kruger.

[53] A listing of Broadband Technology Opportunity Program (BTOP) infrastructure awards made by the National Telecommunications and Information Administration (NTIA) is available at http://www2.ntia.doc.gov/infrastructure. Broadband Initiative Program (BIP) awards made by the Rural Utilities Service (RUS) are available at http://www.rd.usda. gov/files/reports/RBBreportV5ForWeb.pdf.

[54] See http://www.rd.usda.gov/programs-services/farm-bill-broadband-loans-loan-guarantees.

[55] See http://www.rd.usda.gov/programs-services/telecommunications-infrastructure-loans-loan-guarantees.

[56] See http://www.rd.usda.gov/programs-services/community-connect-grants.

[57] Available at http://energycommerce.house.gov/sites/republicans. energycommerce.house.gov/files/114/BILLS-114hrPIH-OpenInternet.pdf.

[58] Amendment offered by Representative Blackburn on House floor, July 15, 2014. The House adopted the amendment by a vote of 223-200 on July 16, 2014. This language was not enacted in the final version of the legislation, P.L. 113- 235.

In: U.S. Broadband Issues
Editor: Chester Rios

ISBN: 978-1-63483-944-0

Chapter 4

COMMUNITY-BASED BROADBAND SOLUTIONS: THE BENEFITS OF COMPETITION AND CHOICE FOR COMMUNITY DEVELOPMENT AND HIGHSPEED INTERNET ACCESS*

The Executive Office of the President

EXECUTIVE SUMMARY

Affordable, reliable access to high speed broadband is critical to U.S. economic growth and competitiveness. Upgrading to higher-speed broadband lets consumers use the Internet in new ways, increases the productivity of American individuals and businesses, and drives innovation throughout the digital ecosystem. As this report describes, while the private sector has made investments to dramatically expand broadband access in the U.S., challenges still remain. Many markets remain unserved or underserved. Others do not benefit from the kind of competition that drives down costs and improves quality. To help fill the void, hundreds of towns and cities around the country have developed their own locally-owned networks. This report describes the benefits of higher-speed broadband access, the current challenges facing the

* This is an edited, reformatted and augmented version of a report issued by the White House, January 2015.

market, and the benefits of competition – including competition from community broadband networks.

~

Since President Obama took office, the United States has significantly expanded its broadband network and increased access. Investments from the federal government have helped deploy or upgrade more than 78,000 miles of network infrastructure since 2009, and more than 45 million Americans have adopted broadband Internet during the President's time in office. Today, more than 90 percent of Americans can access the Internet on a wired line and 98% by either wired or wireless connection.

Competitive markets have helped drive expansion in telecommunications services as strong infrastructure investments and falling prices have opened up a wide range of new communications products and services. Where there is strong competition in broadband markets today, it drives similar improvements. Unfortunately, competition does not extend into every market and its benefits are not evenly distributed. While the U.S. has an extensive network "backbone" of middle-mile connections (long, intra- or interstate physical fiber or cable network connections) with the capacity to offer high-speed Internet to a large majority of Americans, many consumers lack access to the critical "last-mile" (the last legs of the physical network that connect homes and businesses to the broader system), especially in rural areas. It is these last-mile connections that make higher speeds possible. For example, 94 percent of Americans in urban areas can purchase a 25 Mbps (megabit per second) connection, but only 51 percent of the rural population has access to Internet at that speed.

Competition has also been slow to emerge at higher speeds. Nearly forty percent of American households either cannot purchase a fixed 10 Mbps connection (i.e. a wired, land-based connection), or they must buy it from a single provider. And three out of four Americans do not have a choice between providers for Internet at 25 Mbps, the speed increasingly recognized as a baseline to get the full benefits of Internet access.

Without strong competition, providers can (and do) raise prices, delay investments, and provide sub-par quality of service. When faced with limited or nonexistent alternatives, consumers lack negotiating power and are forced to rely on whatever options are available. In these situations, the role of good public policy can and should be to foster competition and increase consumer choice.

At the federal level, the government has already taken active steps to support broadband, committing billions of dollars to deploy middle-mile and last-mile infrastructure, and to ensure that our public schools and libraries have high speed broadband connections.

But local governments also have an important role to play. As this report details, communities around the country like Chattanooga, TN and Wilson, NC have developed a variety of strategies for building locally-owned broadband networks and promoting higher-speed Internet access. Over the past few years, these municipal networks have emerged as a critical tool for increasing access, encouraging competition, fostering consumer choice, and driving local and regional economic development. Local investments have also spurred the private sector to compete for customers, improving services, increasing broadband adoption, and providing more choice for consumers.

Not all communities, however, have the choice to pursue a local broadband network. 19 states currently have barriers in place limiting community broadband and protecting incumbent providers from competition. President Obama believes that there should be a level playing field for community-based solutions and is announcing today a series of steps that the Administration will be taking to foster consumer and community choice.

Economic Benefits of Broadband

In technical terms, broadband refers to a method of transmitting information using many different frequencies, or bandwidths, allowing a network to carry more data. For most Americans, however, the term broadband simply refers to a fast Internet connection—whether fixed or wireless.

Over time, our perceptions of what constitutes a "fast" Internet connection have changed. As consumer and business uses of the Internet evolve, and new applications become more deeply embedded into everyday life, higher speeds frequently shift from being a luxury to a requirement for many users. For example, beginning in 2000 the Federal government defined "broadband" as any service with a download speed of 200 kilobits per second (kbps) or faster.[1] In 2010, the Federal Communications Commission redefined "basic" broadband service as a connection with speeds of at least 4 megabits per second (Mbps) downstream – 20 times faster than the 2000 definition – and at least 1 Mbps upstream.[2]

Today, as everyday experiences for tens of millions of Americans suggest, even these speeds are insufficient for some applications, particularly when a connection is shared by several users. In recognition of the growing need for increased bandwidth, the FCC is considering further revisions to the definition of broadband, and has expressed interest in raising the threshold to 10 or even 25 Mbps downstream and from 1 Mbps to 3 Mbps upstream.[3] The following chart provides a sense of what these definitions mean by showing how long it would take a single user to upload or download different types of content at various connection speeds.

Time Required for Selected Internet-Based Activities at Different Speeds

	3 Minute Song *5 MB (Download)*	**2 Hour Movie** *5 GB (Download)*	**20 Photographs** *40 MB (Upload)*	**5 Minute Video** *200 MB (Upload)*
256 Kbps, 256 Kbps *2000 Broadband*	2m36s	43h24m	20m50s	1h44m
4 Mbps, 1 Mbps *2010 Broadband*	10s	2h46m	5m20s	26m40
25 Mbps, 3 Mbps *Advanced Broadband*	1.6s	26m40s	1m46s	8m53s

Source: CEA Calculations Note: These numbers assume that the ISP is meeting its advertised speed. Download times may be greater during periods of peak traffic.

Demand for Internet access is growing quickly. Total wired and wireless Internet access revenues in 2013 were $140 billion, and have increased by about 15 percent per year in real terms since 2005. [4] The rapidly growing demand for bandwidth is driven by new applications of the Internet that effectively require a broadband connection. These applications, which are increasingly central to everyday life for many Americans, include video streaming, which is used for education, entertainment, and communication; teleworking; cloud storage that allows users to store their files on the Internet, share them, and access them from any device; and online games that allow users to interact with one another in a virtual environment.

Economic studies confirm that broadband Internet creates significant value for consumers and makes an important and rapidly growing contribution to GDP. For example, one study of expenditures for Internet access estimates that as of 2006 – before the widespread availability of streaming audio and video – broadband Internet accounted for $28 billion in U.S. GDP. That study also found that broadband created an additional $5 to $7 billion in consumer surplus in 2006, meaning that consumers would have been willing to pay that much more for the service.[5] Another industry-sponsored study from 2009 estimates that broadband creates $32 billion in annual consumer surplus.[6]

While these studies estimate consumer surplus by examining price sensitivity, another approach is to examine the amount of time users spend online, leading to estimates of $2,500 to $3,800 in value per-user per-year, which imply total consumer surplus in the hundreds of billions of dollars.

Over the longer term, broadband adoption also fuels a virtuous cycle of Internet innovation. This cycle begins when new applications of the Internet create demand for more bandwidth, resulting in a wave of network-level innovation and infrastructure investment. As more bandwidth becomes available, application-sector innovators find new ways to use that capacity, creating additional demand, leading to another round of network investment, and so on. While it is impossible to know what the next bandwidth-hungry killer application will be — perhaps it will be the "Internet of Things" or immersive virtual reality — both history and economic theory show that this virtuous cycle is a powerful driver of innovation and economic growth.[7]

The recent history of wireless broadband provides a good example of the virtuous cycle of innovation and investment. Industry studies suggest that between 2007 and 2011 mobile applications development grew from almost nothing into a $20 billion industry, creating 311,000 U.S. jobs in the process.[8] This led to increased demand for wireless broadband, so that by 2013 private investment in new wireless infrastructure was $34 billion, more than the investments of the big three auto companies combined.[9]

CHALLENGES IN BROADBAND ACCESS AND ADOPTION

Since the President took office, national broadband availability has increased at all advertised speed levels.[10] Today, about 93 percent of Americans have access to wired broadband speeds of at least 3 Mbps downstream (i.e. broadband that allows a user to download 3 megabits per second), and 99 percent of Americans have access to similarly fast mobile wireless broadband. This increased availability reflects both private and public investment, including the $4 billion invested through the National Telecommunications and Information Administration's (NTIA) Broadband Technology Opportunities Program (BTOP) and $3.5 billion invested through the U.S. Department of Agriculture's (USDA) Rural Utilities Service Broadband Initiative Program (BIP), both part of the American Recovery and Reinvestment Act of 2009, as well as $66 million through USDA's ongoing Community Connect grant program.

Share of US with Access to Various Download Speeds, 2013

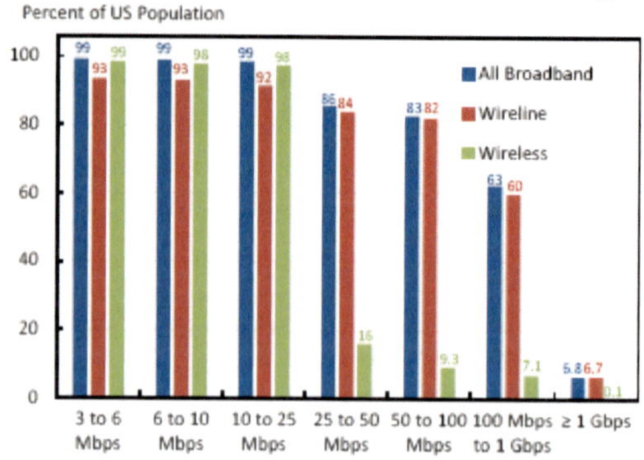

Source: NTIA.

Nevertheless, nearly 51 million Americans cannot purchase a wired broadband connection with download speeds of at least 25 Mbps, and only 63 percent have access to speeds of 100 Mbps or more.[11]

Urban and Rural Communities

One factor that creates disparities in broadband access and adoption is the divide between urban and rural communities. While the gap for the most basic broadband speeds has almost closed (nearly 100 percent of urban residents have access to speeds of 6 Mbps or greater compared to 95 percent of rural residents), rural communities still enjoy far less access to higher speeds. The following figure illustrates this point:

Broadband Availability in Urban and Rural Areas, 2013

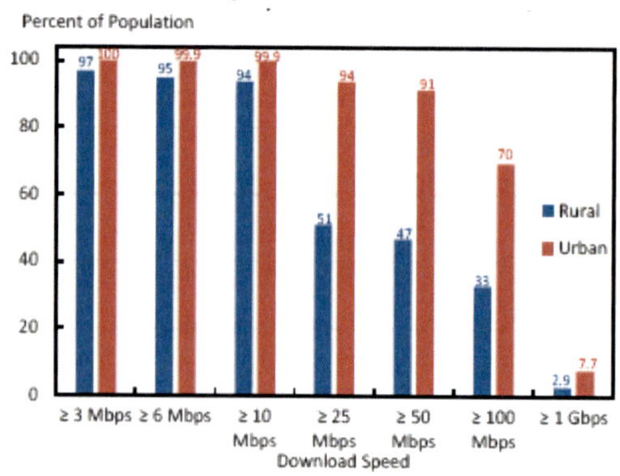

Source: NTIA.

The gap in broadband availability between urban and rural communities is linked to the economics of network investment. The costs of providing a connection increase with distance, and the expected profits increase with the number of customers served. This makes it more economical to serve densely populated urban locations, where shorter wires can serve a larger number of potential customers. While satellite and terrestrial wireless technologies continue to deliver promising improvements, more work is needed to close the urban rural gap in broadband availability.

To address this gap, the USDA, BTOP, and the FCC's Connect America Fund program have all invested in creating the middle-mile infrastructure that provides high-speed access to "anchor institutions" such as schools and libraries in many rural communities. With middle-mile and community infrastructure in place, the remaining challenge is to provide last-mile connections so millions of Americans have access to high-speed broadband. As we describe below, the availability of middle-mile connections creates a significant opportunity for municipalities to increase such access.

Affordability

In total, almost 30 percent of American households did not have a home broadband connection as of 2013. One of the main challenges facing increased broadband adoption is price. In a 2010 survey conducted by the FCC, 36 percent of households without a home broadband connection pointed to expense as the major barrier.[12]

Not surprisingly, the cost of broadband represents a greater obstacle for lower-income Americans than middle- and high-income Americans. The NTIA reports that in 2012, 32 percent of families not online with incomes below $25,000 indicated that the high cost of Internet service prevents them from using broadband at home, compared to less than 22 percent of households not online with annual incomes above $50,000.[13] Overall Internet use is strongly correlated with household income, as illustrated in in the figure below, which plots median income against Internet adoption for a sample of 368 Metropolitan Statistical Areas.

Percent of Households with Internet Access by Metro Area vs. Median Household Income by Metro Area, 2013

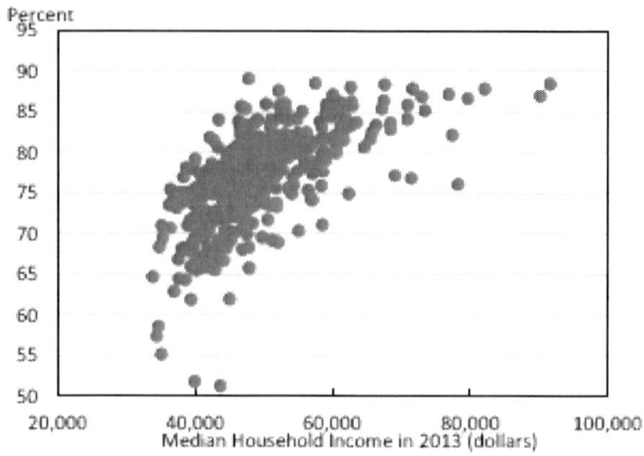

Source: Census.

U.S. broadband is also relatively expensive when compared internationally. The next chart uses data from a recent report on broadband prices in 24 U.S. and international cities.[14] While the 24 cities in this study may not be representative of all urban locations in the U.S. or abroad, it is notable that the median monthly price at each speed level is higher in the U.S., often by 50 percent or more. And while it appears that the U.S. has less price variability at speeds above 75 Mbps, this observation actually reflects the fact that fewer U.S. cities even offer a consumer plan at that level.

Monthy Price of a Yearly Internet Plan by Speed: US vs. World

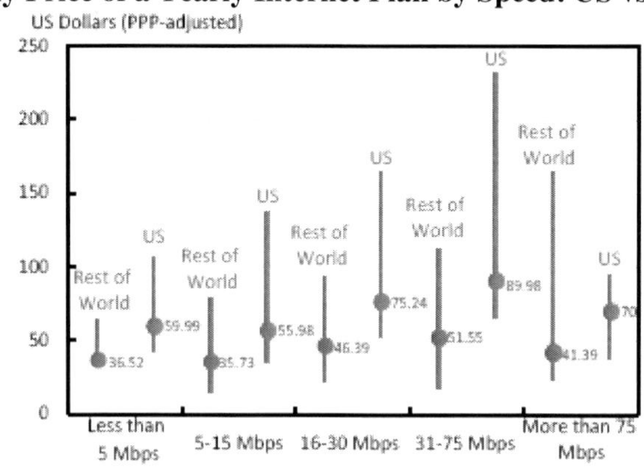

Source: New America Foundation, CEA Calculations.
Note: Lines cover price range, point = median.

BROADBAND COMPETITION

One proven mechanism for increasing Internet access, quality and affordability is to promote competitive markets. Over the past 30 years, telecommunications policy has consistently attempted to encourage market competition in local, long-distance and Internet access markets. For example, the threat of satellite services pushed cable companies to expand their network capacity, positioning them to challenge phone companies in the market for home Internet access. And the ongoing competition between phone and cable companies has created a positive cycle of investment, as providers in many communities continuously upgrade their networks and improve their offerings.[15]

However, the overall national investment picture obscures regional variation. Many local and regional markets today do not have the kind of competition required to continue to ensure affordable access to the higher-speed broadband connections that Americans increasingly require. For example, the following table illustrates the number of choices available to American consumers in fixed and mobile broadband markets. When it comes to wired Internet, which can reliably deliver the highest speeds, the majority of Americans have three choices or less. The situation is somewhat better in wireless markets, although focusing on the number of choices obscures the large share of the market served by a handful of the largest providers. And while competition appears reasonably robust if one focuses on combined choices, it is important to recognize that fixed and wireless Internet are not necessarily substitutes, particularly at speeds of 25 Mbps or higher where there is typically no wireless service available.

Broadband Choice for American Consumers

Number of Choices	Share of U.S. Population (%)		
	Fixed	*Mobile*	*Combined*
1	9	0	0
2	33	3	1
3	37	5	2
4	13	22	4
5	3	26	10
6	1	22	18
7	0	11	19
8+	0	12	46

Source: NTIA, CEA Calculations.

To illustrate the declining level of competition at higher speeds, the following chart shows the number of wired broadband service providers serving American consumers at different speeds. At speeds of 4 Mbps or less, 75 percent of consumers have a choice between two or more fixed providers, and 15 percent can select among three or more ISPs. However, in the market for Internet service that can deliver 25 Mbps downstream – the speed increasingly recognized as a baseline to get the full benefits of Internet access – three out of four Americans do not have a choice between providers.

Number of Choices of Wired Residential Service Provider by Download Speed

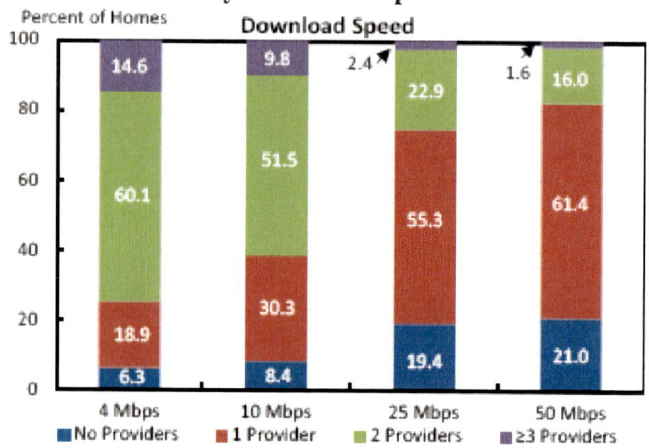

Source: FCC, NTIA.

While increased competition will not necessarily solve all broadband access challenges, basic economics suggests that increased competition leads to a better deal for consumers. For example, a 2014 OECD survey of eleven OECD member countries found that new entrants in wireless markets have a substantial impact on both prices and quality of service. Tellingly, the OECD study indicated that this result occurred even when a market already had three participants – that is, the fourth entrant into a wireless market significantly improved costs and services.[16] As shown above, less than 1 out of 40 American homes has 3 or more choices of providers at speeds in excess of 25 Mbps. Entry also had a positive impact on the market even when the new firm was very small.[17] In the U.S., a 2013 NTIA report found that among those who reported switching their Internet service provider, 38 percent did so to get a better price, and this option is simply unavailable to consumers who are only served by a single Internet Service Provider—or a single provider at the speeds they require.[18]

Even the threat of new competition can lead existing firms to make investments to improve the quality of their goods or services. In the Netherlands, for example, incumbent wireless carriers began offering plans at lower rates in an effort to prevent a new entrant from capturing market share by undercutting existing prices.[1]9 The U.S. cable television industry also provides an example of the benefits of potential competition. Academic research has shown that during the 2000's U.S. cable television operators were more likely to upgrade their systems to allow two-way communications in cities where the cable operator faced a threat of entry from a local municipal electric utility.[20]

Domestic experiences also show how the threat of competition can produce gains for broadband consumers. When Google announced that Google Fiber was coming to Kansas, speeds on existing networks surged 97 percent—the largest year-over-year jump in bandwidth observed in any state, ever. Likewise, when Google indicated that it would begin offering extremely fast connection speeds in Austin, TX, AT&T responded by announcing its own gigabit network.

COMMUNITY-BASED BROADBAND

Where the market does not generate the optimal level of competition or investment, the public sector can step in to make investments, encourage competition and provide choice to consumers. For example, government infrastructure investments, such as those made by the Department of Commerce and Department of Agriculture or by Massachusetts (as described below), may be able to put in place the "middle mile" network that lowers costs of entering the "last mile" market. These investments can attract the private sector or provide local governments the opportunity to build their own systems at much lower prices.

Antitrust and telecommunications policies can also promote competition. At the Federal level, the Department of Justice has an important role to play in preventing the unlawful acquisition or abuse of market power. The Telecommunications Act of 1996 also empowers the FCC to regulate service providers in a manner that promotes competition both within and between technology-based platforms such as cable, cellular, satellite, and wireless. The President's recent call for strong Net Neutrality rules to ensure that no company can act as a gatekeeper to Internet content are fundamentally about preserving access and competition in the digital marketplace. And states have an important role in promoting competition and ensuring fairness in their local communications markets.

But these federal and state initiatives are only part of the solution. Local governments also have a critical role to play. In markets where private competition is anemic, whether because of regulatory barriers to entry or the high fixed costs of infrastructure investment, town and cities can build their own middle-mile networks and offer competitive access to the private sector, as Scott County, MN has done. Or municipalities can provide service directly to consumers, like in Chattanooga, TN. In either case, municipalities are creating more choices for consumers, fostering competition and creating opportunities for economic growth. Municipal broadband is often a logical choice for towns and cities that are already served by a municipal electric utility, since infrastructure costs can be shared across those two services, just as private cable companies leveraged their networks to provide Internet service. Hundreds of towns and cities around the country have experimented with these networks and created tremendous benefits for consumers and businesses. APPENDIX 1 includes a full list of municipal networks around the country.

Today, however, there are barriers to community-owned broadband in 19 states around the country. The Obama Administration believes that consumers should have the option to provide themselves broadband services through local government and locally-owned utilities and that state and local policy should support a level playing field for these community-based solutions. This section considers several detailed case studies of municipal broadband initiatives and their benefits for consumers, businesses and communities.

CHATTANOOGA, TN: GIGABIT SERVICE DRIVES INVESTMENT, INNOVATION

In 2007, Chattanooga's Electric Power Board (EPB), a municipally-owned utility, announced a 10 year plan to build out a fiber network to serve all of Chattanooga. Based on their analysis, EPB had determined that investments in the network could both drive a smart grid system that would generate significant savings by increasing the reliability of its electricity and also provide customers with improved communication services. In 2009, EPB began offering its triple-play services—Internet, phone, and cable television. Since 2009, EPB has upgraded the mid-tier consumer service from 15 to 30, from 30 to 50, and from 50 to 100 Mbps, without raising costs. In 2010, EPB announced it would offer the first 1 gigabit per second (Gbps) service in the United States. Today, EPB operates 8,000 miles of fiber for 60,000 residential and 4,500 business customers out of a potential 160,000 homes and businesses.

EPB's efforts have encouraged other telecom firms to improve their own service. In 2008, for example, Comcast responded to the threat of EPB's entrance into the market by investing $15 million in the area to launch the Xfinity service – offering the service in Chattanooga before it was available in Atlanta, GA. More recently, Comcast has started offering low-cost introductory offers and gift cards to consumers to incentivize service switching. Despite these improvements, on an equivalent service basis, EPB's costs remain significantly lower.

EPB's investments are reshaping Chattanooga's economic landscape. The gigabit broadband service has helped the City attract a new community of computer engineers, tech entrepreneurs and investors. For example, local entrepreneurs have organized Lamp Post, a venture incubator that provides capital and mentorship to startups. Lamp Post now has over 150 employees in a 31,000 square foot office space in downtown Chattanooga. CO.LAB, a local nonprofit organization, provides shared working space, access to investor networks and hosts the annual summer GITANK program, a 14-week business accelerator. The investment community has responded in kind. Since 2009, Chattanooga has gone from close to zero venture capital to at least five organized funds with investable capital of over $50 million. The growing tech ecosystem has been profiled by the *New York Times*, *Washington Post* and *The Atlantic*.

While the broadband network is opening up new economic pathways, EPB itself remains the most important customer for the fiber network, which it has used to develop one of the nation's leading smart grids. The smart grid, which involves 170,000 intelligent electric meters all reporting every 15 minutes, helps EPB monitor and respond to outages, emergencies, and electricity theft in real time. EPB's smart grid has cut duration of power outages by 60 percent, saving local businesses and industry an estimated $45 to $60 million. With the monitoring system in place, EPB crews can also respond in a targeted fashion during emergencies, helping families and businesses cope with tornados and other natural disasters.[21]

WILSON, NC: MUNICIPAL BROADBAND ENCOURAGES PRIVATE COMPETITION

In November of 2006, Wilson's City council voted unanimously to build a fiber-to-thehome (FTTH) network through the town's electricity provider, Greenlight. The City Council issued $28 million in debt to start construction. Greenlight began offering its services in 2008 and expanded its network to include triple-play (television, phone, and internet) services citywide by January 2009. In 2010, the city took another $4.5 million loan from Wells Fargo to improve its network. The subscription base grew steadily in its first few years and numbers over 7000 today –more than a third of Wilson's 21,000 households.

Greenlight has been a commercial success. Greenlight achieved its first monthly operating profit one year ahead of schedule in October 2010 and made a profit of nearly three-quarters of a million dollars in 2013. However, a 2011 state law prevents municipalities from providing broadband service to other towns outside of its area, limiting further growth.

Greenlight's introduction of its triple-play service has increased industry competition, which has lowered prices for Wilson's residents. From 2007 to 2009, Time Warner raised rates for almost all of its services across the board. According to a December 2009 presentation for the House Select Committee on High Speed Internet Access in Rural and Urban Areas, TWC raised rates in non-competitive areas around Wilson while holding Wilson's rates steady. According to the same report, TWC raised its prices for basic internet service in the North Carolina Research Triangle — as much as 52 percent in Cary — but did not impose any rate hike in Wilson. Moreover, TWC stabilized prices in Wilson for the digital sports and games tier, while Triangle customers paid 41 percent more. The lowered prices in Wilson make a big difference. According to an independent consultant for Wilson, Greenlight saved its residents more than $1 million each year compared to what Time Warner Cable customers in other areas pay.

Increased competition has also yielded increased speeds for Wilson customers. Greenlight's system offers speeds of up to 1 gigabit for consumers and businesses. In 2008, Time Warner's residential Road Runner service in the state offered speeds no higher than 10 Mbps, equivalent to Greenlight's lowest consumer tier. TWC charged $57 per month for the service while Greenlight charged $35. In response, TWC upped its top-tier speed to 15 Mbps "because of the competitive environment," according to a Time Warner spokesperson.[22]

LAFAYETTE, LA: NETWORK INCREASES CUSTOMER SAVINGS, STRENGTHENS LOCAL ANCHOR INSTITUTIONS

The residents of Lafayette have a long history of supporting local infrastructure initiatives. Recognizing the need to modernize its broadband infrastructure in the early 2000's, the community voted in 2005 to approve construction of a fiber-to-the-home (FTTH) network. After overcoming serious opposition from local broadband service providers, the publicly-owned Lafayette Utilities System (LUS) started connecting homes and businesses to its LUS Fiber network in 2009. The network seeks to provide equitable access to all of Lafayette's citizens, and the system was rolled out across high-income and low-income neighborhoods equally. LUS Fiber now offers 100 Mbps speed for all subscribers.

As competing firms adjusted their plans to account for LUS Fiber's market entry, residents who weren't customers of the network started to see lower prices. Cox Communications, a major regional provider which had raised rates six times in four years, kept its rates stable from 2004 to 2007 to account for LUS's possible market entry. Still, LUS's prices have been consistently lower than those offered by Cox. Terry Huval, the director of LUS, estimates that the community saved $4 million from these deferred rate increases. Using estimates of Cox's average competing discounts and LUS Fiber's lower rates, LUS projects the fiber system will create total savings of between $90 and $100 million over the its first 10 years.

The fiber network has brought in companies eager to obtain fast service at lower prices. Pixel Magic brought 100 to 200 jobs when it built an office in Lafayette to accomplish work on the movie "Secretariat". The high-speed capability of the broadband network was a big factor in their eventual decision to maintain their office in Louisiana permanently. The tech startup firm Skyscraper Holding moved from Los Angeles to Lafayette to obtain 100 Mb/s speeds at a fraction of the cost the company was charged on the west coast. The company pays just $200 a month for more reliable service.

The network has strengthened community anchors as well, delivering greater value and opportunities for connectivity to Lafayette's school and library systems. By mid-2008, all of the schools in the Lafayette Parish School System were able to access 100 Mbps speeds for $390/month. Not only can students now do more to leverage the Internet for better learning opportunities, this monthly fee saves community tax dollars by being a better value than competitors could offer. Lafayette's public libraries also benefit from the network by sharing a 90 Mbps connection from LUS that was rated as the best value amongst possible providers by the federal E-Rate program.[23]

SCOTT COUNTY, MN: MUNICIPAL GOVERNMENT SEES SAVINGS FOR COUNTY, SCHOOL OPERATIONS

In the early 2000s, Scott County started exploring options for increasing broadband services for county government buildings and schools. In 2007, the County issued $3.5 million in bonds to install a high-speed middle-mile network. The network connects all county-owned facilities, including schools, libraries, city halls, policy and fire departments and public safety towers. It also connects with the state's high capacity backbone network and with multiple private providers. From the beginning, the project was a joint effort between local and state government and the private sector. While the county paid the upfront costs, the state pays for the network's operating costs in exchange for use of the network. The open architecture of the system allows private companies to offer their own services; private providers, in turn, cover the network's maintenance costs.

The network has achieved significant benefits. Scott County's annual bond payment for the construction of the backbone is $35,000 less than what the County was paying for leasing private sector lines. Local schools have seen even greater savings. The costs for Scott County's school districts per megabit of Internet service went from an average of $58.00 to $6.83 per megabit for all school districts—a cost reduction of nearly 90 percent per megabit. The net effect was a tripling of availability (100 to 300 megabits) while costs fell from $5,800 to $2,049 a month. At the state level, the government is saving approximately $1 million per year from access to the public network.

The network has also helped attract significant private investment and fostered job creation. In 2010, for example, Emerson Process Management was finalizing a decision on where to site a new $70 million investment that would create 500 jobs. Emerson's two finalist sites were the town of Shakopee in Scott County, Minnesota and Chihuahua, Mexico. Recognizing the savings from the high-speed broadband network, Emerson chose Scott County.[24]

LEVERETT, MA: STATE AND FEDERAL PROGRAMS ENABLE LOCAL INVESTMENT

In 2008, Massachusetts Governor Deval Patrick created the Massachusetts Broadband Initiative (MBI). MBI was charged with bringing broadband to all residents and businesses in MA within three years. The Broadband Act provided MBI with initial $40 million in state bond funds. Over the last six years, Massachusetts has built 1,200 miles of new fiber optic cable that provide access to more than 120 communities in Western and North Central Massachusetts.

Of the original state funds, $25 million were directed to build a broadband network in Western, MA. With the support of additional federal funds, MBI developed "MassBroadband 123," a middle-mile network serving 123 communities in the region. MBI worked closely with the private sector to build the project. Today, MassBroadband 123 is operated by Axia NGNetworks. The network has an open architecture that allows any Internet service provider to purchase wholesale services on the network at the same rates. The network also positions municipalities to focus on putting homes and businesses on the network through last-mile connections.

Leverett, MA saw the opportunity to build its own broadband system. In 2012, Leverett voters approved a modest property tax increase and a $3.6 million bond to fund the network. Leverett created a publicly controlled Municipal Light Plant (MLP) entity to own and operate its network, named LeverettNet. The town is currently in the process of building the network – which will provide 1 gigabit service – and connecting it to all 630 households in the community.[25]

CHOCTAW NATION TRIBAL AREA, OK: PUBLIC PRIVATE COLLABORATION BRINGS BROADBAND TO NEW COMMUNITIES

In early 2009, much of the ten Southeastern Oklahoma counties encompassed by the Choctaw Nation's Tribal Area lacked access to reliable broadband service. The low population density (8.3 to 19.7 people per square mile), the high poverty rate (25 percent of the population below the poverty line) and the rugged terrain made the economics of broadband infrastructure very challenging. Initial capital costs to deploy broadband meant that broadband service was limited only to commercially viable areas.

Pine Tele, the service provider offering voice, video, cell, long distance, and high-speed broadband in SE OK applied for and received 4 American Recovery and Reinvestment awards in 2009 and 2010. One grant was to build out fiber to the home in the area already covered by landlines, and the other three were for wireless – advanced 3G technology – to completely unserved areas. As of September 2014 Pine Tele had deployed 324 miles of fiber, 5,500 fiber drops, and 54 tower sites. New or improved broadband service had been made available to 1,757 fiber customers and 1,194 wireless customers. Today, Pine Telephone provides a variety of broadband packages over both their fiber and wireless facilities ranging from 1.5 Mbps to 5 Mbps for download speeds and 384 Kbps to 5 Mbps for upload speeds.

The benefits for the community have been significant. Every school in the 10 county Pine Tele service area is now connected with high-speed fiber optic broadband service. This has created the ability to integrate online educational tools into everyday teaching and assessments of student comprehension. Broken Bow School District is one example. This district serves approximately 1,280 students per day. They have been able to integrate smart boards, iPads, online lesson plans, and the "I-Ready program" to supplement learning. Hundreds of performance tests are now completed online. And family engagement is improved, as parents are increasingly provided online access to records of attendance, assignments, and test scores. The connectivity also allows the Choctaw Nation to multicast educational videos and share messages from Tribal leadership from a central location. For example, the Choctaw School of Language now offers distance learning courses to approximately 14 head starts and 32 high schools within the Choctaw Nation, in addition to several universities.[26]

Promoting Broadband that Works

Last November, the President outlined his plan to keep the Internet open to new competition and innovation by safeguarding net neutrality — which will help ensure no one company can act as a gatekeeper to digital content. But there is more work to do so that every American has access to a free and open internet. This is particularly true in areas where broadband competition is lacking, resulting in high prices and slow service.

High-speed, low-cost broadband is paving the way for economic revitalization not just in Cedar Falls, but in places like Chattanooga, TN and Lafayette, LA — which have Internet speeds up to 100 times faster than the national average and deliver it at an affordable price. To help more communities achieve these results, support economic growth, and promote a level playing field for all competitors, the Obama Administration is:

- *Calling to End Laws that Harm Broadband Service Competition:* Laws in 19 states — some specifically written by special interests trying to stifle new competitors — have held back broadband access and, with it, economic opportunity. Today President Obama is announcing a new effort to support local choice in broadband, formally opposing measures that limit the range of options to available to communities to spur expanded local broadband infrastructure, including ownership of networks. As a first step, the Administration is filing a letter with the Federal Communications Commission (FCC) urging it to join this effort by addressing barriers inhibiting local communities from responding to the broadband needs of their citizens.
- *Expanding the National Movement of Local Leaders for Better Broadband:* As of today, 50 cities representing over 20 million Americans have joined the Next Century Cities coalition, a nonpartisan network pledging to bring fast, community-supported broadband to their towns and cities. They join 37 research universities around the country that formed the Gig.U partnership to bring fast broadband to communities around their campuses. To recognize these remarkable individuals and the partnerships they have built, in June 2015 the White House will host a Community Broadband Summit of mayors and county commissioners from around the nation who are joining this movement for broadband solutions and economic revitalization.
- *Announcing a New Initiative to Support Community Broadband Projects:* To advance this important work, the Department of Commerce is launching a new initiative, BroadbandUSA, to promote broadband deployment and adoption. Building on expertise gained from overseeing the $4.7 billion Broadband Technology Opportunities Program funded through the Recovery Act, BroadbandUSA will offer online and in-person technical assistance to communities; host a series of regional workshops around the country;

and publish guides and tools that provide communities with proven solutions to address problems in broadband infrastructure planning, financing, construction, and operations across many types of business models.

- *Unveiling New Grant and Loan Opportunities for Rural Providers:* The Department of Agriculture is accepting applications to its Community Connect broadband grant program and will reopen a revamped broadband loan program which offers financing to eligible rural carriers that invest in bringing high-speed broadband to unserved and underserved rural areas.
- *Removing Regulatory Barriers and Improving Investment Incentives:* The President is calling for the Federal Government to remove all unnecessary regulatory and policy barriers to broadband build-out and competition, and is establishing a new Broadband Opportunity Council of over a dozen government agencies with the singular goal of speeding up broadband deployment and promoting adoptions for our citizens. The Council will also solicit public comment on unnecessary regulatory barriers and opportunities to promote greater coordination with the aim of addressing those within its scope.

APPENDIX 1: U.S. MUNICIPALITIES WITH BROADBAND NETWORKS[27]

City	State	Name of Network	Type
Ketchikan	AK	KPU Telecommunications	cable
Kotlik	AK	Kotlik	cable
Statewide	AK	Rural Alaska Video E-Health Network (RAVEN)	inet
White Mountain	AK	White Mountain	cable
Opelika	AL	Opelika	fiber
Opp	AL	Opp Cablevision	cable
Scottsboro	AL	Scottsboro EPB	cable
Sylacauga	AL	Sylacauga	cable
Conway	AR	Conway Corporation	cable
Paragould	AR	Paragould Light Water and Cable	cable
Sells	AZ	Tohono O'odham Last-Mile FTTH and Broadband Wireless Network	partial
Anaheim	CA	Anaheim	dark
Anaheim	CA	Anaheim Fiber	inet

City	State	Name of Network	Type
Burbank	CA	Burbank Water and Power	partial
Glendale	CA	Glendale	dark
Humboldt County	CA	Digital Redwoods	inet
Loma Linda	CA	Loma Linda	dark
Loma Linda	CA	Loma Linda Connected Community	fiber
Lompoc	CA	City of Lompoc (LompocNet)	inet
Long Beach	CA	Long Beach	dark
Mendocino County	CA	Mendocino Community Network	inet
Palo Alto	CA	Palo Alto Fiber	dark
Pasadena	CA	Pasadena	dark
San Bruno	CA	San Bruno Municipal Cable TV	cable
San Francisco	CA	SF Fiber	question
Santa Clara	CA	Santa Clara	partial
Santa Monica	CA	Santa Monica City Net	partial
Santa Monica	CA	Santa Monica Fiber	partial
Shafter	CA	City of Shafter, California	partial
Truckee	CA	Truckee Donner Public Utility District	dark
Vernon	CA	Vernon Light & Power	fiber
Cortez	CO	Cortez Community Network	partial
Durango	CO	Durango	dark
Glenwood Springs	CO	Glenwood Springs Community Broadband Network (GSCBN)	partial
Longmont	CO	NextLight	fiber
Bristol	CT	Bristol CT	inet
East Hartford	CT	Connecticut Education Network	dark
Manchester	CT	Manchester Wireless	inet
Fort Pierce	FL	FPUAnet Communications	partial
Gainesville	FL	GATOR NET	partial
Hobe Sound	FL	Martin County Dark Fiber	dark
Indiantown	FL	Martin County Dark Fiber	dark
Jacksonville	FL	Jacksonville iNet	inet
Jensen Beach	FL	Martin County Dark Fiber	dark
Jupiter Island	FL	Martin County Dark Fiber	dark
Lakeland	FL	Lakeland	dark
Leesburg	FL	Leesburg	partial
New Smyrna Beach	FL	Utilities Commission, City of New Smyrna Beach	inet
Ocala	FL	Ocala Utility Services	partial
Ocean Breeze Park	FL	Martin County Dark Fiber	dark
Palm Beach County	FL	Palm Beach County	partial
Palm City	FL	Martin County Dark Fiber	dark
Palm Coast	FL	Palm Coast FiberNET	partial

Appendix 1. (Continued)

City	State	Name of Network	Type
Port Salerno	FL	Martin County Dark Fiber	dark
Quincy	FL	NetQuincy	fiber
Sewall's Point	FL	Martin County Dark Fiber	dark
Stuart	FL	Martin County Dark Fiber	dark
Tallahassee	FL	Tallahassee	dark
Valparaiso	FL	Valparaiso Broadband	cable
Baconton	GA	Community Network Services - Camilla	cable
Baker County	GA	SGRITA Rural Last-mile Infrastructure Project Last-mile	partial
Cairo	GA	Community Network Services - Cairo (Syrup City)	cable
Calhoun	GA	CALNET	partial
Calhoun County	GA	SGRITA Rural Last-mile Infrastructure Project Last-mile	partial
Camilla	GA	Community Network Services - Camilla	cable
Cartersville	GA	Fibercom	partial
Catoosa County	GA	OptiLink	partial
Columbia County	GA	Columbia County Community Broadband Network	partial
Dalton	GA	OptiLink	fiber
Doerun	GA	City of Doerun	cable
Douglasville	GA	Douglas County School System Fiber	inet
Dublin	GA	Dublin	partial
Early County	GA	SGRITA Rural Last-mile Infrastructure Project Last-mile	partial
Elberton	GA	Elberton Utilities	cable
Flintstone	GA	EPB Fiber Optics	fiber
Forsyth	GA	Forsyth Cablenet	cable
LaGrange	GA	LaGrange Telecommunications Department	partial
Miller County	GA	SGRITA Rural Last-mile Infrastructure Project Last-mile	partial
Mitchell County	GA	SGRITA Rural Last-mile Infrastructure Project Last-mile	partial
Monroe	GA	Monroe Utilities Network	cable
Moultrie	GA	Community Network Services - Moultrie	cable
Murray County	GA	OptiLink	partial
Pelham	GA	Community Network Services - Pelham (Pelnet)	cable
Rossville	GA	EPB Fiber Optics	fiber
Sandersville	GA	Sandersville FiberLink	partial
Thomasville	GA	Community Network Services - Thomasville	cable

City	State	Name of Network	Type
Tifton	GA	Tifton	dark
Whitfield County	GA	OptiLink	partial
Wildwood	GA	EPB Fiber Optics	fiber
Algona	IA	Algona Municipal Utilities	cable
Alta	IA	Altatec	cable
Bellevue	IA	Bellevue	fiber
Cedar Falls	IA	Cedar Falls Utilities	fiber
Cedar Falls	IA	Cedar Falls Utilities - rural expansion	partial
Coon Rapids	IA	Coon Rapids Municipal Utilities	cable
Grundy Center	IA	Grundy Center Municipal Light & Power	cable
Harlan	IA	Harlan Municipal Utilities	cable
Hartley	IA	The Community Agency	cable
Hawarden	IA	HITEC - Hawarden Integrated Technology, Energy, & Communication	cable
Independence	IA	Independence Light & Power, Telecommunications	cable
Indianola	IA	Indianola	partial
Laurens	IA	Laurens Municipal Power and Communications	cable
Lenox	IA	Lenox	fiber
Manning	IA	Manning Municipal Communication and Television System Utility	cable
Mapleton	IA	Mapleton Communications	cable
Muscatine	IA	MachLink	cable
Osage	IA	Osage Municipal Utilities	cable
Paullina	IA	The Community Agency	cable
Primghar	IA	The Community Agency	cable
Reinbeck	IA	Reinbeck Telecom	cable
Sanborn	IA	The Community Agency	cable
Spencer	IA	Spencer Municipal Utilities	fiber
Webster City	IA	Webster City	dark
Ammon	ID	Ammon	partial
Idaho Falls	ID	Circa	dark
Plummer	ID	Coeur d'Alene Reservation FTTH Project Last-mile Non-remote	partial
Aurora	IL	Onlight Aurora	partial
Aurora	IL	OnLight Aurora	dark
Champaign	IL	Urbana-Champaign Big Broadband UC2B	partial
DeKalb County	IL	DeKalb Advancement of Technology Authority Broadband	partial
Evanston	IL	Evanston	partial
Highland	IL	Highland Communication Services	fiber
LaSalle County	IL	DeKalb Advancement of Technology Authority Broadband	partial

Appendix 1. (Continued)

City	State	Name of Network	Type
Princeton	IL	Princeton Municipal Utilities	partial
Rochelle	IL	Rochelle Municipal Utilities	partial
Rock Falls	IL	Rock Falls	partial
Urbana	IL	Urbana-Champaign Big Broadband UC2B	partial
Anderson	IN	Anderson Municipal Light and Power	partial
Auburn	IN	Auburn Essential Services	fiber
Lebanon	IN	Lebanon Utilities	cable
Mishawaka	IN	Saint Joe Valley MetroNet	dark
South Bend	IN	Saint Joe Valley MetroNet	dark
Westfield	IN	City of Westfield	partial
Chanute	KS	Chanute	partial
Lenexa	KS	Lenexa Fiber	dark
Ottawa	KS	Ottawa Network	partial
White Cloud	KS	Iowa Tribe of Kansas and Nebraska Fiberto-the-Premise	partial
Barbourville	KY	Barbourville	cable
Bardstown	KY	Bardstown Cable	cable
Bowling Green	KY	Bowling Green Municipal Utility	partial
Corinth	KY	City of Williamstown	partial
Frankfort	KY	Frankfort Plant Board	cable
Franklin	KY	Franklin Municipal FiberNET	partial
Glasgow	KY	Glasgow Electric Power Board	cable
Grant County	KY	City of Williamstown	partial
Hopkinsville	KY	Energy Net	cable
Monticello	KY	Community Telecom Services	cable
Murray	KY	Murray Electric System	cable
Owen County	KY	City of Williamstown	partial
Owensboro	KY	OMU Online	partial
Paducah	KY	Paducah Power System	partial
Russellville	KY	Russellville EPB SmartNet	fiber
Williamstown	KY	City of Williamstown	cable
Lafayette	LA	Lafayette Utilities System	fiber
Braintree	MA	Braintree Electric Light Department	cable
Chicopee	MA	Chicopee Electric Light	partial
Holyoke	MA	Holyoke Gas & Electric Co.	partial
Leverett	MA	LeverettNet	fiber
Norwood	MA	Norwood Light Broadband	cable
Russell	MA	Russell Municipal Cable	cable
Shrewsbury	MA	Shrewsbury Electric and Cable Operations	cable
South Hadley	MA	Five College Fiber Optic Network	inet

City	State	Name of Network	Type
Taunton	MA	Taunton Municipal Lightning Plant	partial
Worcester	MA	Worcester Municipal Fiber Loop	inet
Carroll County	MD	Carroll County Broadband	dark
Columbia	MD	Howard County Fiber Network	dark
Dayton	MD	Howard County Fiber Network	dark
Easton	MD	EastonOnline	cable
Elkridge	MD	Howard County Fiber Network	dark
Ellicot City	MD	Howard County Fiber Network	dark
Fulton	MD	Howard County Fiber Network	dark
Highland	MD	Howard County Fiber Network	dark
Savage	MD	Howard County Fiber Network	dark
Coldwater	MI	CBPU	cable
Crystal Falls	MI	City of Crystal Falls	cable
Holland	MI	Holland Fiber Network	fiber
Negaunee	MI	City of Negaunee Dept. of Public Works	cable
Norway	MI	City of Norway CATV System	cable
Sebewaing	MI	Sebewaing Light & Water	fiber
Wyandotte	MI	Wyandotte	cable
Bagley	MN	Bagley Public Utilities	fiber
Barnesville	MN	Barnesville Municipal Utilities	partial
Belle Plaine	MN	Scott County Fiber Network	dark
Bingham Lake	MN	SMBS - Bingham lake	fiber
Brewster	MN	SMBS - Brewster	fiber
Carver	MN	CarverLink	dark
Chanhassen	MN	CarverLink	dark
Chaska	MN	Chaska.Net	partial
Cologne	MN	CarverLink	dark
Crosslake	MN	Crosslake Communications	fiber
Eagan	MN	Access Eagan	partial
Elko New Market	MN	Scott County Fiber Network	dark
Hamburg	MN	CarverLink	dark
Heron Lake	MN	SMBS - Heron Lake	fiber
Jackson	MN	SMBS - Jackson	fiber
Jordan	MN	Scott County Fiber Network	dark
Lake County	MN	Lake County	partial
Lakefield	MN	SMBS - Lakefield	fiber
Mayer	MN	CarverLink	dark
Monticello	MN	Monticello Fiber Network	fiber
New Germany	MN	CarverLink	dark
New Prague	MN	Scott County Fiber Network	dark
Norwood Young America	MN	CarverLink	dark

Appendix 1. (Continued)

City	State	Name of Network	Type
Okabena	MN	SMBS – Okabena	fiber
Pine City	MN	Pine City Fiber Optic Backbone	partial
Prior Lake	MN	Scott County Fiber Network	dark
Round Lake	MN	SMBS - Round Lake	fiber
Savage	MN	Scott County Fiber Network	dark
Shakopee	MN	Scott County Fiber Network	dark
Silver Bay	MN	Lake County Fiber Network	partial
St. Louis Park	MN	St. Louis Park	inet
Two Harbors	MN	Lake County Fiber Network	partial
Victoria	MN	CarverLink	dark
Waconia	MN	CarverLink	dark
Watertown	MN	CarverLink	dark
Westbrook	MN	Westbrook Municipal Light & Power	cable
Wilder	MN	SMBS - Wilder	fiber
Windom	MN	Windomnet	fiber
Kahoka	MO	Kahoka	cable
Marshall	MO	Marshall	fiber
North Kansas City	MO	liNKCity	fiber
Poplar Bluff	MO	City of Poplar Bluff Municipal Utilities	cable
Springfield	MO	SpringNet	partial
Collins	MS	Collins Communications	cable
Asheville	NC	ERC Broadband	dark
Chapel Hill	NC	Chapel Hill Fiber Optic Services	inet
Cornelius	NC	MI-Connection	cable
Davidson	NC	MI-Connection	cable
Mooresville	NC	MI-Connection	cable
Morganton	NC	Morganton	cable
Salisbury	NC	Fibrant	fiber
Sylva	NC	BalsamWest FiberNET	partial
Tryon	NC	PANGAEA	partial
Wilson	NC	Greenlight	fiber
South Sioux City	NE	South Sioux City Municipal Network	inet
Cheshire	NH	Fast Roads	dark
Claremont	NH	Fast Roads	dark
Enfield	NH	Fast Roads	partial
Fitzwilliam	NH	Fast Roads	dark
Goshen	NH	Fast Roads	dark
Hanover	NH	Fast Roads	dark
Keene	NH	Fast Roads	dark
Lebanon	NH	Fast Roads	dark

City	State	Name of Network	Type
Lyme	NH	Fast Roads	dark
Marlow	NH	Fast Roads	dark
New London	NH	Fast Roads	dark
Newport	NH	Fast Roads	dark
Orford	NH	Fast Roads	dark
Richmond	NH	Fast Roads	dark
Rindge	NH	Fast Roads	partial
Springfield	NH	Fast Roads	dark
Sunapee	NH	Fast Roads	dark
Swanzey	NH	Fast Roads	dark
Glassboro	NJ	Glassboro Municipal Area Network	inet
Vineland	NJ	Vineland Metropolitan Area Network	inet
Churchill	NV	CC Communications	fiber
Bristol Center	NY	Axcess Ontario	dark
Bristol Springs	NY	Axcess Ontario	dark
Canandaigua	NY	Axcess Ontario	dark
Cheshire	NY	Axcess Ontario	dark
Clifton Springs	NY	Axcess Ontario	dark
East Bloomfield	NY	Axcess Ontario	dark
Farmington	NY	Axcess Ontario	dark
Fishers	NY	Axcess Ontario	dark
Geneva	NY	Axcess Ontario	dark
Gorham	NY	Axcess Ontario	dark
Hogansburg	NY	St. Regis Mohawk Tribe Connect (Economic Development for the 21st Century)	partial
Honeoye	NY	Axcess Ontario	dark
Hopewell	NY	Axcess Ontario	dark
Manchester	NY	Axcess Ontario	dark
Naples	NY	Axcess Ontario	dark
New York City	NY	New York City Wireless Network NYCWiN	inet
Phelps	NY	Axcess Ontario	dark
Rushville	NY	Axcess Ontario	dark
Stanley	NY	Axcess Ontario	dark
Victor	NY	Axcess Ontario	dark
West Bloomfield	NY	Axcess Ontario	dark
Akron	OH	OneCommunity	partial
Ashtabula	OH	OneCommunity	partial
Barberton	OH	OneCommunity	partial
Bryan	OH	Bryan Municipal Utilities	cable
Butler County	OH	Butler County	inet
Canton	OH	OneCommunity	partial
Cincinnati	OH	Hamilton County	inet

Appendix 1. (Continued)

City	State	Name of Network	Type
Cleveland	OH	OneCommunity	partial
Cleveland Heights	OH	OneCommunity	partial
Dover	OH	Dover Technology	dark
Dublin	OH	Dublink+	partial
Eastlake	OH	OneCommunity	partial
Elyria	OH	OneCommunity	partial
Gahanna	OH	Gahanna	inet
Hamilton	OH	Hamilton Miami U	inet
Lorain	OH	OneCommunity	partial
Mayfield Village	OH	OneCommunity - Mayfield Village	partial
Medina County	OH	Medina County	dark
Mentor	OH	OneCommunity	partial
Middletown	OH	Middletown Miami U	inet
New Albany	OH	BlueAlbany	partial
Sandusky	OH	OneCommunity	partial
Wadsworth	OH	City of Wadsworth Electric & Communications Dept.	cable
Wadsworth	OH	OneCommunity	dark
Woodsfield	OH	Woodsfield Municipal Power	cable
Wooster	OH	OneCommunity	partial
Ponca City	OK	Ponca City Technology Services	partial
Sallisaw	OK	DiamondNet	fiber
Ashland	OR	Ashland Fiber Network	cable
Canby	OR	Clackamas Broadband Express	dark
Damascus	OR	Clackamas Broadband Express	dark
Douglas County	OR	Oregon South Central Regional Fiber Consortium Lighting the Fiber Middle-mile Project	partial
Estacada	OR	Clackamas Broadband Express	dark
Eugene	OR	Eugene	dark
Gladstone	OR	Clackamas Broadband Express	dark
Government Camp	OR	Clackamas Broadband Express	dark
Happy Valley	OR	Clackamas Broadband Express	dark
Independence	OR	MINET	fiber
Klamath County	OR	Oregon South Central Regional Fiber Consortium Lighting the Fiber Middle-mile Project	partial
Lane County	OR	Oregon South Central Regional Fiber Consortium Lighting the Fiber Middle-mile Project	partial

City	State	Name of Network	Type
Milwaukie	OR	Clackamas Broadband Express	dark
Molalla	OR	Clackamas Broadband Express	dark
Monmouth	OR	MINET	fiber
Mulino	OR	Clackamas Broadband Express	dark
Oregon City	OR	Clackamas Broadband Express	dark
Sandy	OR	SandyNet	partial
Sherwood	OR	Sherwood Fiber	partial
Springfield	OR	Springfield Utility Board	dark
The Dalles	OR	Q-Life Network	partial
Wilsonville	OR	Clackamas Broadband Express	dark
Beaver County	PA	Beaver County Fiber	inet
Kutztown	PA	Hometown Utilicom	fiber
Pitcairn	PA	Pitcairn Power/Community Cable	cable
Hartsville	SC	Hartsville	question
Oconee County	SC	Oconee FOCUS (Fiber Optics Creating Unified Solutions)	partial
Orangeburg County	SC	Orangeburg	partial
Aberdeen	SD	CityNet (Dakota Interconnect)	inet
Beresford	SD	Beresford Municipal Telephone/Cablevision	cable
Brookings	SD	Swiftel	fiber
Bristol	TN	Bristol TN Essential Services	fiber
Chattanooga	TN	EPB Fiber Optics	fiber
Clarksville	TN	Clarksville CDE Lightband	fiber
Columbia	TN	CPWS Broadband	cable
East Ridge	TN	EPB Fiber Optics	fiber
Erwin	TN	Erwin Utilities	partial
Fayetteville	TN	Fayetteville Public Utilities	cable
Jackson	TN	Jackson Energy Authority	fiber
Johnson City	TN	BVU OptiNet	partial
Lookout Mountain	TN	EPB Fiber Optics	fiber
Morristown	TN	FiberNET	fiber
Nashville	TN	NESNet	dark
Pulaski	TN	PES Energize	fiber
Red Bank	TN	EPB Fiber Optics	fiber
Ridgeside	TN	EPB Fiber Optics	fiber
Signal Mountain	TN	EPB Fiber Optics	fiber
Tullahoma	TN	Tullahoma Utilities Board	fiber
Greenville	TX	GEUS	cable
Lindon	UT	Utah Telecommunications Open Infrastructure Agency (UTOPIA)	partial
Brigham City	UT	Utah Telecommunications Open Infrastructure Agency (UTOPIA)	fiber

Appendix 1. (Continued)

City	State	Name of Network	Type
Centerville	UT	Utah Telecommunications Open Infrastructure Agency (UTOPIA) HQ	fiber
Layton	UT	Utah Telecommunications Open Infrastructure Agency (UTOPIA)	partial
Midvale	UT	Utah Telecommunications Open Infrastructure Agency (UTOPIA)	partial
Murray	UT	Utah Telecommunications Open Infrastructure Agency (UTOPIA)	partial
Orem	UT	Utah Telecommunications Open Infrastructure Agency (UTOPIA)	partial
Payson	UT	Utah Telecommunications Open Infrastructure Agency (UTOPIA)	partial
Perry	UT	Utah Telecommunications Open Infrastructure Agency (UTOPIA)	partial
Spanish Fork	UT	Spanish Fork Community Network	cable
Tremonton	UT	Utah Telecommunications Open Infrastructure Agency (UTOPIA)	fiber
West Valley City	UT	Utah Telecommunications Open Infrastructure Agency (UTOPIA) HQ	partial
Abingdon	VA	BVU OptiNet	fiber
Arlington County	VA	ConnectArlington	dark
Atkins	VA	BVU OptiNet	partial
Bluefield	VA	BVU OptiNet	partial
Bristol	VA	BVU OptiNet	fiber
Castlewood	VA	BVU OptiNet	partial
Cedar Bluff	VA	BVU OptiNet	partial
Chillhowie	VA	BVU OptiNet	partial
Clay Pool Hill	VA	BVU OptiNet	partial
Cleveland	VA	BVU OptiNet	partial
Clinchco	VA	BVU OptiNet	partial
Clintwood	VA	BVU OptiNet	partial
Damascus	VA	BVU OptiNet	partial
Danville	VA	nDanville	partial
Duffield	VA	LENOWISCO Planning District Commission	partial
Eastern Virginia	VA	Eastern Shore of Virginia Broadband Authority	question
Emery-Meadow View	VA	BVU OptiNet	partial
Galax	VA	Wired Road	partial
Glad Spring	VA	BVU OptiNet	partial
Grundy	VA	BVU OptiNet	partial
Haysi	VA	BVU OptiNet	partial

City	State	Name of Network	Type
Hiltons	VA	BVU OptiNet	fiber
Honaker	VA	BVU OptiNet	partial
Independence	VA	BVU OptiNet	partial
Lebanon	VA	BVU OptiNet	partial
Luray	VA	Page County Broadband Project	partial
Marion	VA	BVU OptiNet	partial
Martinsville	VA	Martinsville Information Network - MINET	partial
Nelson County	VA	Nelson County Virginia Broadband Project	partial
Page County	VA	Page County Broadband Project	partial
Richlands	VA	BVU OptiNet	partial
Rockbridge County	VA	Connect the Dots: Rockbridge Broadband Initiative	partial
Rural Retreat	VA	BVU OptiNet	partial
Saltville	VA	BVU OptiNet	partial
Shenandoah	VA	Page County Broadband Project	partial
St Paul	VA	BVU OptiNet	partial
Stanley	VA	Page County Broadband Project	partial
Staunton	VA	Staunton	dark
Sugar Grove	VA	BVU OptiNet	partial
Tazewell	VA	BVU OptiNet	partial
Troutdale	VA	BVU OptiNet	partial
Vansant	VA	BVU OptiNet	partial
Wytheville	VA	BVU OptiNet	partial
Barnard	VT	ECFibernet (East Central Vermont Community Fiber Network)	partial
Bethel	VT	ECFibernet (East Central Vermont Community Fiber Network)	partial
Braintree	VT	ECFibernet (East Central Vermont Community Fiber Network)	partial
Brookfield	VT	ECFibernet (East Central Vermont Community Fiber Network)	partial
Hancock	VT	ECFibernet (East Central Vermont Community Fiber Network)	dark
North Randolph	VT	ECFibernet (East Central Vermont Community Fiber Network)	partial
Pomfret	VT	ECFibernet (East Central Vermont Community Fiber Network)	partial
Reading	VT	ECFibernet (East Central Vermont Community Fiber Network)	dark
Rochester	VT	ECFibernet (East Central Vermont Community Fiber Network)	dark
Royalton	VT	ECFibernet (East Central Vermont Community Fiber Network)	partial

Appendix 1. (Continued)

City	State	Name of Network	Type
Sharon	VT	ECFibernet (East Central Vermont Community Fiber Network)	partial
Stockbridge	VT	ECFibernet (East Central Vermont Community Fiber Network)	dark
Aberdeen	WA	Grays Harbor PUD	partial
Ardenvoir	WA	Chelan PUD	partial
Bauer's Landing	WA	Douglas County Community Network	inet
Benton City	WA	Benton PUD Broadband	partial
Benton County	WA	Benton PUD Broadband	partial
Blewett	WA	Chelan PUD	partial
Bridgeport	WA	Douglas County Community Network	inet
Bridgeport Bar	WA	Douglas County Community Network	inet
Burlington	WA	Mt Vernon Fiber Optic Services	partial
Cashmere	WA	Chelan PUD	fiber
Chelan	WA	Chelan PUD	partial
Chelan County	WA	Chelan PUD	fiber
Cheney	WA	Cheney Fiber Network	partial
Chumstick	WA	Chelan PUD	partial
Clallam County	WA	Clallam PUD	partial
Coulee City	WA	Grant PUD	partial
Coulee Dam	WA	Grant PUD	partial
Desert Aire	WA	Grant PUD	fiber
Desert Canyon	WA	Douglas County Community Network	inet
Douglas County	WA	Douglas County Community Network	inet
Dryden	WA	Chelan PUD	fiber
East Wenatchee	WA	Douglas County Community Network	inet
Edmonds	WA	City of Edmonds	dark
Entiat	WA	Chelan PUD	partial
Ephrata	WA	Grant PUD	partial
Franklin County	WA	Franklin PUD Broadband	partial
Grand Coulee	WA	Grant PUD	fiber
Grant County	WA	Grant PUD	fiber
Hartline	WA	Grant PUD	fiber
Kennewick	WA	Benton PUD Broadband	fiber
Kitsap County	WA	Kitsap PUD	fiber
Leavenworth	WA	Chelan PUD	fiber
Mansfield	WA	Douglas County Community Network	inet
Mason County	WA	Mason County PUD3	partial
Mattawa	WA	Grant PUD	fiber
Meritt	WA	Chelan PUD	partial

City	State	Name of Network	Type
Monitor	WA	Chelan PUD	fiber
Moses Lake	WA	Grant PUD	partial
Mt Vernon	WA	Mt Vernon Fiber Optic Services	partial
Newport	WA	Pend Oreille County Public Utility District (PUD) Broadband Network	partial
Okanogan County	WA	Okanogan PUD	fiber
Orondo	WA	Douglas County Community Network	inet
Pacific County	WA	Pacific County PUD#2	partial
Pasco	WA	Franklin PUD Broadband	fiber
Pend Oreille County	WA	Pend Oreille PUD	fiber
Peshastin	WA	Chelan PUD	fiber
Port of Skagit County	WA	Mt Vernon Fiber Optic Services	partial
Prosser	WA	Benton PUD Broadband	fiber
Quincy	WA	Grant PUD	fiber
Royal City	WA	Grant PUD	fiber
Sequim	WA	Clallam PUD	partial
Shelton	WA	Mason County Public Utilities District	partial
Soap Lake	WA	Grant PUD	fiber
Sun Cove	WA	Douglas County Community Network	inet
Tacoma	WA	Click! Network	cable
Warden	WA	Grant PUD	fiber
Waterville	WA	Douglas County Community Network	inet
Wenatchee	WA	Chelan PUD	fiber
Wilson Creek	WA	Grant PUD	fiber
Yodelin	WA	Chelan PUD	partial
Eau Claire	WI	Chippewa Internetworking Consortium (CINC)	inet
Oconto	WI	Oconto Falls Municipal Utilities	cable
Platteville	WI	Chippewa Internetworking Consortium (CINC)	partial
Reedsburg	WI	Reedsburg Utility Commission	fiber
Reedsburg	WI	Reedsburg Utility Commission - rural expansion	partial
Shawano	WI	Shawano Municipal Utilities	fiber
Sun Prairie	WI	Sun Prairie Utilities	partial
Superior	WI	Chippewa Internetworking Consortium (CINC)	partial
Wausau	WI	Chippewa Internetworking Consortium (CINC)	partial
Philippi	WV	Philippi Communications System	fiber
Powell	WY	Powell Fiber Optic Network	fiber

End Notes

[1] National Telecommunications and Information Administration (NTIA). 2013. "US Broadband Availability: June 2010-June 2012." http://www.ntia.doc.gov/files/ntia/publications/usbb_avail_report_05102013.pdf, 3.

[2] Federal Communications Commission. 2010. "Sixth Broadband Deployment Report." July 20. https://apps.fcc.gov/edocs_public/attachmatch/FCC-10-129A1_Rcd.pdf.

[3] Federal Communications Commission. 2014. "Tenth Broadband Progress Notice of Inquiry." August 5. https://apps.fcc.gov/edocs_public/attachmatch/FCC-14-113A1.pdf.

[4] United States Census Bureau. "Table 4: Estimated Sources of Revenue for Employer Firms: 2010 through 2013." November 19. Annual and Quarterly Services. https:// www.census.gov/services/index.html (CEA Calculations).

[5] Greenstein, Shane and Ryan C. McDevitt. 2009. "The Broadband Bonus: Accounting for Broadband Internet's Impact on U.S. GDP" NBER Working Paper No. 14758. http://www.nber.org/papers/w14758. This study suggests that roughly 50 percent of this amount– $8 to $11 billion– was "new GDP" that would not have been realized but for the replacement of dial-up with broadband connections.

[6] Dutz, Mark, Jonathan Orszag, and Robert Willig. 2009. "The Substantial Consumer Benefits of Broadband Connectivity for U.S. Households." COMPASS LEXECON, commissioned by the Internet Innovation Alliance, July. http://internetinnovation.org/files/special-reports/CONSUMER_BENEFITS_OF_BROADBAND.pdf.

[7] Bresnahan, Timothy F., and Shane Greenstein. 1999. "Technology Competition and the Structure of the Computer Industry." Journal of Industrial Economics 47: 1-40, DOI: 10.1111/1467-6451.00088.

[8] Mandel, Dr. Michael. 2012. "Where the Jobs Are: The App Economy." February 2. https://www.conference-board.org/retrievefile.cfm?filename=Where-the-Jobs-Are_-The-AppEconomy.pdf&type=subsite.

[9] US Telecom. The Broadband Association. "Broadband Investment." http://www.ustelecom.org/broadband-industry/broadband-industry-stats/investment; Carew, Diana G. and Michael Mandel. 2012. "Investment Heroes: Who's Betting on America's Future?" Progressive Policy Institute, July. http://progressivepolicy.org/wp-content/uploads/2012/07/07.2012-Mandel_Carew_Investment-Heroes_Whos-Betting-on-Americas-Future.pdf.

[10] NTIA, 2013.

[11] Office of Science and Technology Policy & The National Economic Council. 2013. "Four Years of Broadband Growth." June. http://www.whitehouse.gov/sites/default/files/broadband_report_final.pdf.

[12] Horrigan, John B. 2010. "Broadband Adoption & Use in America. Results from an FCC Survey.". March. http://transition.fcc.gov/DiversityFAC/032410/consumer-surveyhorrigan.pdf.

[13] Department of Commerce (NTIA and ESA). 2013. "Exploring the Digital Nation. America's Emerging Online Experience." June. http://www.ntia.doc.gov/files/ntia/publications/exploring_the_digital_nation_- _americas_emerging_online_experience.pdf, page 39.

[14] Hibah, Hussain, Danielle Kehl, Patrick Lucey, and Nick Russo. 2013. "The Cost of Connectivity 2013. Data Release: A comparison of high-speed Internet prices in 24 cities around the world." New America Foundation, October. http://oti.newamerica.net/publications/policy/the_cost_of_connectivity_2013. The US cities included are: Bristol (VA), Chattanooga (TN), Kansas City (MO), Lafayette (LA), Los Angeles (CA), New York (NY), San Francisco (CA), and Washington (DC). The international cities included are:

Toronto (Canada), Hong Kong, Prague (Czech Republic), Copenhagen (Denmark), Paris (France), Berlin (Germany), Dublin (Ireland), Tokyo (Japan), Riga (Latvia), Mexico City (Mexico), Amsterdam (Netherlands), Bucharest (Romania), Seoul (South Korea), Zurich (Switzerland), and London (UK). Some of these cities are omitted from the data underlying the chart below, as the plans included in NAF's dataset are not all annual contracts.

[15] Wheeler, Tom. 2014. "Prepared Remarks of Chairman Tom Wheeler. 'The Facts and Future of Broadband Competition.'" September 4. https://apps.fcc.gov/edocs_public/attachmatch/DOC329161A1.pdf.

[16] OECD. 2014. "Wireless Market Structures and Network Sharing." OECD Digital Economy Papers, No. 243. OECD Publishing. http://dx.doi.org/10.1787/5jxt46dzl9r2-en.

[17] The case of 3UK is cited by OECD as particularly indicative of this reality.

[18] Department of Commerce (NTIA and ESA). 2013. "Exploring the Digital Nation. America's Emerging Online Experience." June. http://www.ntia.doc.gov/files/ntia/publications/exploring_the_digital_nation_-_americas_emerging_online_experience.pdf, page 39.

[19] OECD. 2014.

[20] Seamans, Robert. 2012. Fighting City Hall: Entry Deterrence and Technology Deployment in the Cable TV Industry. Management Science 58(3): 461-475.

[21] Sources used to prepare this case study: Technology, Gig, and Entrepreneurship Task Force. 2014. "Chattanooga Forward." March 19. http://www.chattanooga.gov/images/Mayor_Images/Chattanooga_Forward_Working_report. pdf; Wyatt, Edward. 2014. "Fast Internet is Chattanooga's New Locomotive." New York Times, February 3. http://www.nytimes.com/2014/02/04/technology/fast-internet-service-speedsbusiness-development-in-chattanooga.html; Fung, Brian. 2013. "How Chattanooga beat Google Fiber by half a decade." Washington Post, September 17. http://www.washingtonpost.com/blogs/the-switch/wp/2013/09/17/how-chattanooga-beatgoogle-fiber-by-half-a-decade; Rushe, Dominic. 2014. "Chattanooga's Gig: how one city's superfast internet is driving a tech boom." The Guardian, August 30. http://www.theguardian.com/world/2014/aug/30/chattanooga-gig-high-speed-internet-techboom; Cosco, Joey. 2014. "A City in Tennessee Has The Big Cable Companies Terrified." Business Insider, July 30. http://www.businessinsider.com/chattanooga-tennessee-biginternet-companies-terrified-2014-7; Mitchell, Christopher. 2012. "Broadband At the Speed of Light. How Three Communities Built Next-Generation Networks." April. Institute for Local Self-Reliance and Benton Foundation, April. http://ilsr.org/wp-content/uploads/2012/04/muni-bbspeed-light.pdf.

[22] Sources used to prepare this case study:
Mitchell, Christopher. O'Boyle, Todd. 2012. "Carolina's Connected Community: Wilson Gives Greenlight to Fast Internet." December. Institute for Local Self-Reliance and Common Cause Education Fund, December. http://ilsr.org/wp-content/uploads/2012/12/wilsongreenlight.pdf; New York Times. 2014. "Communities Fight State Laws That Can Divide Broadband Access." November 9. http://www.nytimes.com/2014/11/10/technology/in-ruralamerica-challenging-a-roadblock-to-high-speed-internet.html; Indy Week. 2008. "Mighty, mighty broadband." June 18. http://www. indyweek.com/indyweek/mighty-mightybroadband/Content?oid=1209049; Wired.com. 2011. "North Carolina Enacts Pro-ISP, Anti-Municipal Broadband Law." May 23. http:// www.wired.com/2011/05/nc-gov-anti-munibroadband/; Davidson, Charles M. Santorelli, Michael J. 2014. "Understanding The Debate Over Government-owned Broadband Networks: Context, Lessons Learned, and a Way Forward for Policy Makers." June. Advanced Communications Law and Policy Institute at New York Law School, June. http://www.nyls.edu/advanced-communications-law-and-policy-institute/wp-content/uploads/

sites/169/2013/08/ACLP-%E2%80%93-Wilson-Case-Study-%E2%80%93- June-2014.pdf; Wilson Times. 2010. "Greenlight competition affects rates elsewhere." http://www. wilsontimes.com/greenlight/Story/Greenlight-competition-affects-rates-elsewhere23Sources used to prepare this case study:

Huval, Terry. 2010. Testimony Before the Committee on Small Business and Entrepreneurship. Hearing on "Connecting Main Street to the World: Federal Efforts to Expand Small Business Internet Access." Senate Hearing 111-1166, April 27. http://www. sbc.senate.gov/public/index.cfm?a=Files.Serve&File_id=646b01b6-6e75-4f5a-9c0f-790c0ba48889.

Jervis, Rick. 2012. "Louisiana city blazes high-speed Web trail." USA Today, February 5. http://usatoday30.usatoday.com/news/nation/story/2012-02-01/broadband-telecomlafayette/52920278/1

Mitchell, Christopher. 2012. "Broadband at the Speed of Light: How Three Communities Built Next-Generation Networks." April. Institute for Local Self-Reliance and Benton Foundation, April. http://ilsr.org/wp-content/uploads/2012/04/muni-bb-speed-light.pdf

Pollick, Michael. 2010. "Dark lines are a draw for business." Sarasota Harold-Tribune, September 7. http://www.heraldtribune.com/article/20100907/ARTICLE/100909825/-1/news?p=all&tc=pgall

[24] Sources used for this case study: Collaborative Governance Council. 2010. "Scott County Broadband (Fiber) Project." Minnesota Office of the State Auditor, August 12. http://www.osa.state.mn.us/Other/councils/CollaborativeGovernance/081210/ScottCountyB ro adbandProjectPresentation.pdf; Minnesota Ultra High Speed Broadband Task Force. 2009. "Scott County Broadband (Fiber) Approach." February 20. http://www.ultra-high-speedmn.org/CM/Custom/Shelton%20-%20Broadband%20Task%20force%20 Presentation.pdf; Moore, Janet. 2013. "Scott County approves incentives for Shutterfly." Minnesota Star Tribune, August 20. http://www.startribune.com/blogs/220393971.html; Mitchell, Chris, and Lisa Gonzalez. 2014. "All Hangs on Deck. Minnesota Local Government Models for Expanding Fiber Internet Access." Institute for Local Self-Reliance, September. http://www.ilsr.org/wpcontent/uploads/downloads/2014/09/ all_ hands_on_deck_mn.pdf.

[25] Sources used for this case: Gonzalez, Lisa. 2012. "Network Moves Forward in Leverett, Western Massachusetts." Community Broadband Networks, December 20. http://www. muninetworks.org/content/network-moves-forward-leverett-westernmassachusetts;

Crawford, Susan and Robyn Mohr. 2013. "Bringing Municipal HighSpeed Internet Access to Leverett, Massachusetts." Harvard University Berkman Center Research Publication No. 26, December 17. http://papers.ssrn.com/sol3/papers.cfm?abstract_id=2366044; Lennett, Benjamin, Patrick Lucey, Joanne Hovis, and Andrew Afflerbach. "The Art of the Possible: An Overview of Public Broadband Options." New American Foundation Open Technology Institute and CTC Technology & Energy, May 6. http://newamerica.net/sites/newamerica. net/files/policydocs/TheArtofthePossibleOverviewof PublicBroadbandOptions_NAFOTI-CTC.pdf; Scherban, Debra. 2014. "Crocker Communications to provide Internet Service for Leverett's own broadband network." Daily Hampshire Gazette, September 2. http://www.gazettenet.com/home/13387776-95/crocker-communications-toprovide-internet-service-for-leveretts-own-broadband-network.

[26] Sources used for this case: Yocum, Katie. 2010. "USDA Recovery Act-Funded Wireless Internet to Reach Remote Choctaw Nation." United States Department of Agriculture, June 11. http://blogs.usda.gov/2010/06/11/usda-recovery-act-funded-wireless-Internet-to-reach-remote-choctaw-nation; The Choctaw Nation of Oklahoma. "Programs & Services for the

Choctaw People." http://s3.amazonaws.com/choctawmsldigital/assets/2345/choctaw_services_original.pdf. Material from a site visit conducted by USDA from September 22 to 25, 2014 in Broken Bow, Oklahoma was also relied upon for the Pine Telephone Company Case Study.

[27] The Institute for Local Self-Reliance, data provided to NEC/CEA upon request.

INDEX

C

D

E

F

G

H

I

J

K

L

M

N

O

P

T

U